Modern Management of Water and Wastewater Utilities

WATER MANAGEMENT SERIES

Series Editor: Gordon L. Culp

Handbook of Wastewater Collection and Treatment:
Principles and Practice
M. ANIS AL-LAYLA
SHAMIM AHMAD
E. JOE MIDDLEBROOKS

Handbook of Biological Wastewater Treatment:
Evaluation, Performance, and Cost
HENRY H. BENJES, JR.

Handbook of Sludge-Handling Processes:
Cost and Performance
GORDON L. CULP

Handbook of Industrial Wastes Pretreatment
JON C. DYER
ARNOLD S. VERNICK
HOWARD D. FEILER

Modern Management of Water and Wastewater Utilities
WILLIAM E. KORBITZ

Modern Management of Water and Wastewater Utilities

William E. Korbitz

WATER MANAGEMENT SERIES

Garland STPM Press
New York & London

The book is dedicated to my wife, Dee, without whose inspiration, encouragement, and assistance the book could not have been written.

Library of Congress Cataloging in Publication Data
Korbitz, William E., 1926-
    Modern management of water and wastewater utilities.
    (Water management series)
    Includes index.
    1. Water-supply—Management.   2. Sewage disposal—
Management.   I. Title.   II. Series.
TD353.K618   363.6'1'068   81-2162
ISBN 0-8240-7069-0   AACR2

Published by Garland STPM Press
136 Madison Avenue, New York, New York 10016

Printed in the United States of America

15   14   13   12   11   10   9   8   7   6   5   4   3   2   1

Water Management Series

The Garland series in water management addresses the many aspects of water resource management which must be carefully integrated to protect the quality and quantity of our water supplies. The series is a comprehensive compilation of volumes dealing with the most important aspects of water resource management. Topics span the range from detailed information on design, cost, and performance of specific types of treatment processes to the environmental implications associated with certain management approaches. The series is designed to be a timely and authoritative reference source for practicing engineers and planners, as well as for regulatory agencies and students.

Series Editor:
GORDON L. CULP

Contents

16 / Research and Development   **249**

Preface

The management of water and wastewater utilities has become increasingly complex during recent years. This increased complexity has resulted in a need for instructional and reference materials for all of the operations and maintenance, engineering, planning, financial, legal, and public relations aspects of water and wastewater utility management. The enactment of several federal laws and the promulgation of scores of implementation regulations have placed many new constraints on the managers of water and wastewater utilities. These constraints include requirements pertaining to planning, construction grants, discharge permits, sludge handling and disposal, treatment levels, public participation, and monitoring. Many books are available on such individual topics as water and wastewater technology, concepts of business administration and public administration, public and utility financing, public relations, and statute and contract law, but there is a dearth of sources which bring together the many aspects of modern day management of water and wastewater utilities.

It is the primary purpose of this book to fill this gap and to serve as a ready reference to utility managers for practical approaches to the many problems which confront them. This book also is intended as text to be used in classes on water and wastewater utility management for students in engineering, public administration and public works administration. The book should also be a helpful reference for consultants, for federal and state enforcement personnel, for city councils and utility boards of directors, and for students of engineering and public administration who have a special interest in water and wastewater utilities.

Modern Management of Water and Wastewater Utilities is based largely on the personal experience of the author in the broad field of public works, mainly in the field of wastewater utility management. Managerial concepts for wastewater utilities are much like those for water utilities and many of the treatment operations are similar, though the water and wastewater treatment processes as a whole are different. The distribution of water in a public water supply system is similar in many respects to wastewater collection and transportation, even though water supply systems are pressure systems, while wastewater collection and transportation operate mostly by gravity systems. The use of customer service charges rather than taxation for the majority of

xiii

capital improvements and operations and maintenance expenses of both water and wastewater utilities makes the management of the two utilities even more similar.

The encouragement of the members of my family and of many friends and working associates during the writing of this book is gratefully acknowledged. Also to be acknowledged are my past employers, especially the Board of Directors of the Metropolitan Denver Sewage Disposal District No. 1, who provided me many opportunities to develop and implement the procedures, concepts, and detailed programs which are the basis of this book. A special acknowledgement is due my wife Dee, who not only provided the ultimate in encouragement and inspiration throughout the writing of this book but also typed all the drafts and provided other valuable assistance.

1

Management Responsibilities and Constraints

The management of water and wastewater utilities has over a period of many years grown and developed from the relatively simple, day to day supervision of operations and maintenance activities by technical people, supported by a system of financing based on special assessments and property taxes, into a complex system of coordinated technology, finance, planning, public relations, politics, laws, and regulations. The relatively few simple technological, financial, legal, and political constraints within which water and wastewater utility managers have worked have likewise become increasingly sophisticated, and have been joined by a panoply of rather complex and conflicting sociological, environmental, public relations, legislative, and regulatory constraints.

Utility Manager Responsibilities

The responsibilities of the water or wastewater utility manager are no longer limited to successful performance of those duties required to build, operate, and maintain suitable water or wastewater facilities. At the turn of the century the water utility manager was responsible for locating and developing a water supply source, transporting water to a place of treatment, and then treating and distributing the water to the customers of the system. In the case of public water supplies which emanated from underground aquifers through wells the task was even simpler. The water utility manager experienced relatively few constraints and was subject to virtually no control from federal or state government. The wastewater utility manager similarly was responsible for building, operating, and maintaining a comparatively simple

1

sewer system, which transported wastewater from homes, businesses, and industries to a waterway for disposal.

As the pollution of the waterways gave rise to legitimate concern it became necessary for the wastewater utility manager to provide for the construction, operation, and maintenance of a primary treatment plant and the disposal of the small quantities of sludge which resulted from treatment. Progressively more stringent laws and regulations have placed increasing burdens on managers of water and wastewater utilities to provide better-quality water to customers of public water systems and higher-quality treatment of wastewater for protection of the public health and the environment. These responsibilities of water and wastewater utility managers have increased in complexity and scope, especially as a result of an increasing number of laws, regulations, and permit conditions, and in response to protests, questions raised in environmental assessments, and the concerns of federal and state legislators and regulatory agencies.

The basic responsibility of the manager of a water or wastewater utility is to provide a specific water or wastewater service to the customers of the utility. The service to be provided by the water utility to the public, and for which the water utility manager is responsible, is the furnishing of water which is safe, palatable, and in adequate quantity, and which is reliably delivered at adequate pressure. The water utility is expected to provide the public with water which in addition to meeting current requirements for protection of public health, is also free of objectionable taste or odor, and does not contain significant color or turbidity. The water distribution system must be capable of providing enough water for the many uses within and around residences, by commercial customers, and by industry and institutions. These required quantities of water must be provided at adequate pressure at all times to not only meet the daily needs of the various customers, but also to provide for satisfactory fire protection. The water supply must be reliable even when the utility experiences interruptions caused by water main breaks, by unusually heavy demands by industry or other large users, by unusual weather conditions such as severe cold weather, and by fire fighting activities. The water utility manager, in addition to this basic responsibility to his customers, also is responsible for providing water service in accordance with applicable laws, regulations, political pressures, environmental demands, available revenues, and the desires of the public.

The wastewater utility manager has a similar responsibility: to collect wastewaters from homes, businesses, industries, and institutions; to transport the wastewater to a treatment facility; to treat the wastewater to suitable standards before discharging it to a waterway; and to dispose of the residues of treatment in a satisfactory manner. As with the public water supply, it is necessary for wastewater collection, transportation, and treatment to be reliable and uninterrupted. The sanitary sewer system must be capable of transporting wastewater from the individual customers to the point of treatment, as much as possible by gravity, with virtually no threat of disruption of flow. The treatment facilities must be so designed, constructed, operated, and maintained that, on a continuing, uninterrupted basis, the wastewater receives the degree of treatment required for protecting the quality of water in the receiving waterway, and for meeting the treatment standards dictated by federal and state laws and regulations. The wastewater utility manager thus has the responsibility to ensure that the required service is provided to the customers of the wastewater utility without interruption, and that the service is provided economically and in accordance with the many conflicting demands of political, environmental, and numerous other special interest groups.

The responsibilities of water and wastewater utility management can be summarized as the protection of public health and welfare; the reasonable protection of the environment; the furnishing of water and wastewater service which is adequate, reliable, and safe; the improvement of service whenever and however possible while at the same time minimizing the costs to the consumer; the maintaining of a healthy fiscal, political, and public relations atmosphere for the utility; and, of ever increasing importance, the conservation of resources. In the area of resource conservation, it is essential for managers of water and wastewater utilities to not only conserve energy, water, chemicals, and materials, but also to develop resource recovery programs and new sources of usable resources.

The Manager's Constraints

The water and wastewater utility manager is faced with many constraints within which he is required to perform his managerial duties and meet his management responsibilities. These constraints include federal and state laws and regulations; limitations on water supply quality and quantity; limitations imposed on wastewater treatment activities because of existing or proposed use of water from the waterway into which treated effluent is discharged; technological limitations in process design or operation of facilities; laboratory technology; available equipment; finances; energy availability; political implications; environmental and sociological concerns; water rights; and the influence of collective bargaining and fair employment activities. Discussion of these constraints in some detail in this chapter and in the following chapters will provide guidance to the utility manager in his coping with these and other similar constraints.

Regulatory Constraints

Federal and state laws did not constitute a serious constraint to either water or wastewater utility management until the advent of the significant environmental awareness of the general public in the 1950s and 1960s. This environmental awareness, based largely on rhetoric and emotion rather than on knowledge or true concern on the part of citizens, led to congressional enactment of stringent laws which were intended to protect the environment by rigidly regulating all aspects of wastewater treatment, and also to protect public health by rigidly regulating the management and operation of public water supply facilities. The federal laws were not serious constraints until the 1970s, when the Federal Water Pollution Control Act Amendments of 1972, the Clean Water Act of 1977 and the Safe Drinking Water Act, all containing stringent constraints and requirements, were enacted by Congress. The Federal Water Pollution Control Act Amendments of 1972 and the Clean Water Act of 1977 were intended to provide for improved water quality in the nation's lakes, oceans, and streams to the extent that the nation's waterways would safely support water recreation and an increased number of forms of aquatic life, but with much less apparent concern for public health. The Safe Drinking Water Act, which was enacted by Congress in 1974, required drastic changes in planning, designing, and operating the facilities involved in the treatment of public water supplies. Many of these changes resulted from a certain level of hysteria about the causes of health defects, in particular cancer.

Most of the states also replaced relatively weak water pollution control laws and regulations with rigid water quality standards, stream classifications, and discharge permit conditions, in accordance with the requirements of the Federal Water Pollution Control Act. The strict control of public water supply, treatment, and distribution, along with the rigid control of wastewater collection, transmission, treatment, and residue disposition, without any realistic regard for total costs and real benefits to the public, will continue to provide challenging constraints to the managers of water and wastewater utilities.

The implementation of many provisions of federal and state laws has resulted in many specific constraints and requirements not envisioned by the legislators, and often in regulations which are contrary to the intent of the law. Many of these regulations relate to the manner in which the managers of water and wastewater utilities are required to perform their duties, for example, in the following areas: the identification and establishment of levels and types of water and wastewater treatment; the amount and types of construction with accompanying uncontrolled costs; various types and levels of planning; the methods and extent of facilities planning, financing, and charging methods; and voluminous environmental impact statements which have been costly and time-consuming, resulting in delays in much urgently needed construction.

Some of these regulations and their effects will be discussed in some detail in a later chapter; a brief discussion is, however, appropriate here. The planning required of wastewater utilities by the Federal Water Pollution

Control Act amendments of 1972 includes state, regional, basin, and local facilities planning. The regulations have required information, reviews, and approvals of little or no value to the operating agencies, which have resulted in significant delays in completing the local facilities and fiscal planning, and consequent delays in implementation. The delays in construction have not only caused local agencies to continue the difficult and costly operation of inadequate facilities until new facilities were constructed and placed in operation, but have also resulted in increased construction costs. The environmental impact statement procedure, which was a key part of the implementation of the National Environmental Policy Act of 1969, has likewise required voluminous documents based on extensive studies, some of which have borne little or no relationship to the water pollution problem being studied. The documents are not easily understood, contain vast amounts of irrelevant information, and have generally required unreasonable periods of time for review and issuance in final form.

The laws and implementing regulations are desirable in a general sense, but rigid adherence to specified implementation procedures has imposed additional constraints on responsible utility management and has been an impediment to progress.

Physical Constraints

Limited raw water sources for public water supplies are a serious constraint in water utility management, especially for the western and southwestern states. This water supply limitation often leads to the location and development of increasingly more expensive sources of supply; the construction of expensive transmission, storage, and treatment facilities; the occasional imposition and enforcement of water use restrictions; and the development and implementation of extensive water conservation and reuse programs. Although the total water resources available throughout the country may be adequate for the indefinite future, the concentration of the bulk of people in urban areas makes it necessary to likewise concentrate water of suitable quality in those same urban areas. The development of water shortages in certain urban areas has imposed constraints on managers of both water and wastewater utilities which had not been encountered before.

Under water shortage conditions the water utility manager has found it necessary to plan not only for more expensive water sources, but also for new types of sources, such as the exchange of water with other users or the direct processing and use of wastewater treatment plant effluent. Similarly, the wastewater utility manager is faced under water shortage conditions with new decisions concerning the location and type of wastewater treatment, which are governed by the location and method of final discharge or use of the treatment plant effluent. Years ago, the effluent was merely discharged into a receiving stream for ultimate disposal; now the effluent may be routed to another treatment facility for further use in a potable water system, or it may require a different type of total treatment for agricultural use, or a higher level or different type of treatment for meeting water quality standards which are intended to permit fishing and water recreation use of the receiving waterway.

Technological Constraints

Technological constraints of both major and minor significance have historically posed problems for water and wastewater utility managers only for short periods of time. Research and development efforts in the field have always resulted in new or improved technology which removed the technological constraints. The appropriate water or wastewater utility officials were then convinced of the merits of the new technology, at which time the necessary funds were provided by the utility officials for implementing the new technology.

As federal and state regulatory agencies exercised increasingly rigid control over the water or wastewater treatment provided by water and wastewater utilities, numerous and lengthy delays were encountered in securing approval from the regulatory agencies for the use of newly developed technology. Planning for implementation of the new technology and the acquisition of the necessary funds followed soon after approval by the regulatory agencies, but additional delays were experienced because of additional grant or permit conditions imposed by the same agencies. Another technological constraint which must be faced by water and wastewater utility managers is caused by industrial activities. The development of new industrial processes with accompanying new industrial wastes requires the development of new methods to remove or neutralize the industrial waste contaminants by either water treatment or wastewater treatment.

Even with the delays and problems caused by the regulations which limit the selection of treatment alternatives, water and wastewater utility managers have usually been able to successfully meet the technological challenges and surmount the technological constraints which have confronted them. If water and wastewater utilities are to provide adequate service to their customers and protect public health and the environment, the utility managers must develop the capability of successfully dealing with any technological constraints with which they may be confronted in the future.

Fiscal Constraints

As has been true since the first water and wastewater systems were established, financial limitations are and will continue to seriously constrain water and wastewater utility managers. Federal and state requirements for increased levels of treatment of both water and wastewater tend to conflict with growing opposition by the public to increased expenditures and taxation, and make fiscal planning difficult at best. For many years in the early days of water utilities, the utility manager would plan far ahead to ensure that his utility would be able to secure, transport, treat, and distribute to the public reasonably clean and safe water; he determined how to raise the funds to finance specific water facility projects, and also provided the expertise to design and build the projects. Local water utility officials, based on the advice of their manager, determined when new facilities were needed, what the new facilities should be, and how they should be financed. The utilities seemed almost always to be able to provide a type and level of service to the public which the public needed, could afford, and were willing to finance. The same was also true in the wastewater field. Here, too, local wastewater utility officials determined the level of wastewater treatment which was needed to protect public health, aquatic life, and the general environment; weighed the cost of that level of treatment against available funds and the probable benefits; and proceeded with the necessary projects.

With the coming of the era of the so-called environmentalists, new requirements for types and levels of additional water and wastewater treatment increased the treatment costs to the point where orderly and logical fiscal planning became virtually impossible because of the total financial burden being borne by the consumer. Influences from outside the water or wastewater utility service area have made enormous fiscal demands on the utilities and their customers, so that in the future constraints on available funds will require the increasing attention of utility managers.

Energy Constraints

With a serious energy shortage facing most of the world, it has become essential that the future treatment and transmission of water and wastewater be accomplished with a minimum consumption of energy. To do this it will be necessary for operating procedures as well as the planning and design of transmission and treatment facilities to be continually evaluated and revised. To successfully cope with the energy shortage it is also necessary for managers of water and wastewater utilities to find ways to operate their facilities with a smaller quantity of chemicals, less water, and less of other resources as well, especially of those resources whose production, transportation, and use require large amounts of energy. These managers and their personnel will also have to search for ways to develop the production of new or replacement physical resources and new energy supplies. It is more important than ever, for example, for the design of sewer systems to favor gravity flow systems over systems which require the use of pump stations; the usual economic comparison of the higher capital cost of gravity flow with the total capital, operation, and maintenance cost of pumping now include an evaluation of energy consumption as well. Formulation of decisions concerning the location and design of treatment facilities now also must include a greater concern for the impact on power requirements and other resource requirements than was shown in the days when cost was the major consideration in the evaluation of alternatives. Likewise, the selection of treatment process alternatives, such as for wastewater treatment plant sludge processing and disposal, must include an increased consideration of the quantities and types of chemicals and energy-intensive processes and selection of resource recovery processes rather than resource consumptive processes. In addition, consideration of the probability that natural gas, electric power, certain chemicals, and other resources may not be available in adequate quantities, may be too costly, or may not be available at all will increasingly govern each alternative evaluation as future resource shortages occur.

Legal Constraints

In the western states the use of water has historically been restricted by water rights laws. These laws were necessary because of the inadequate quantities of rainfall and runoff available to meet the needs of agriculture and then, following urbanization of many areas, the needs of urban residential, commercial, and industrial water users as well. For many years water rights have been considered nothing less than sacred and it has cost water and wastewater utilities considerable time, effort, and money to keep from infringing on them. The rights of agricultural water users or other water users to certain stream flows which would otherwise be available for urban use has often made it necessary for urban water supplies to be developed far from urban areas, usually at much greater cost than if the legal rights of the other users could have been disregarded.

Water utility managers also have been legally prohibited from reusing water after it has left their systems. In addition, the development of properly coordinated water supply–wastewater treatment systems which could provide considerable benefits to the customers of both systems has at times been delayed or even prevented by concern about the rights of other agencies or individuals to the water. These are but a few examples of legal constraints caused by the western water laws and the water rights of water users. Many other types of legal constraints impact the activities and effectiveness of water and wastewater utility managers; some of these will be discussed in a later chapter.

Personnel Constraints

With an increase in labor union representation in water and wastewater utilities it has become necessary for utility managers to devote an increasing amount of their time to contract negotiations, grievances and arbitration, strike planning, and the many other accompanying fiscal and personnel problems. These factors, together with increased involvement of federal legislation in matters of employment, have made personnel matters which previously were handled with relative ease extremely complex and the subject of much negotiation, conflict, and legislation. The motivation of individual employees toward improved productivity has become difficult because collective bargaining has removed much of the opportunity for employees to receive awards or other recognition for outstanding or above standard performance of duties. The usual grievance procedure makes heavy demands on the time of management personnel for settling questions which previously were settled with relatively little expenditure of time or effort.

A major problem caused by union organization and representation of employees is the lack of flexibility in using employees interchangeably for any of a number of jobs. This also tends to prohibit an employee from developing new skills because of contractual limitations on the types of work specific employees may perform in certain positions or job classifications. The need to comply with federal regulations dealing with equal employment opportunity, employment quotas, and the like have placed serious constraints on utility managers in their efforts to retain highly qualified employees who can provide satisfactory service without incurring an unreasonable increase in cost.

Political Constraints

Although problems with political patronage and political interference with management of water and wastewater utilities have diminished over the years, it is still important for utility managers to understand and be capable of working effectively within political influences and within various political systems. It is reasonable to expect that as the activities of water and wastewater utilities are increasingly governed by regulatory and legislative controls, by the activities of labor unions, environmental and other special interest groups, and by more effective news media activities, the problems caused by political influences will be of less concern to the utility manager in the future. It will remain important, however, for the manager of a water or wastewater utility to understand the various political systems within which he must perform his management duties and to develop the ability to work effectively within the political constraints forced on him.

Public Relations Constraints

Various constraints which may be classified as environmental and sociological can cause considerable difficulty for water and wastewater managers. As the general public has become increasingly aware of environmental problems and concerns, they have begun to notice things which previously had gone unobserved and unheeded. Foam on a river downstream from the discharge of a wastewater treatment plant, odors coming from wastewater treatment or sludge disposal, dust from a construction site, and noises from any type of facility all are now noticed by and cause complaints from the public. Many phenomena which for many years had been an integral part of the life of everyone now have become the basis of objection, complaint, conflict, and legal action. Employees as well as the general public are continually reminded of their many rights and privileges, and are encouraged to resist any activity or situation which they feel may infringe on those rights and privileges. Such minor occurrences as traffic congestion, water or sewer service interruptions, water or sewer service rate increases, extra long work days, safety concerns, and many other such matters cause serious complaints to be showered upon the managers of water and wastewater utilities. Every one of these complaints must be thoughtfully considered and thoroughly investigated so that satisfactory answers and solutions can be provided.

Communications Constraints

A constraint which appears to become increasingly serious each year is the inability of most people to adequately communicate in both written and verbal form. This is as true of water and wastewater utility employees as with the general public, and warrants concern on the part of utility managers. Concern has been voiced about the relatively high percentage of high school graduates who cannot read and write, but little concern has been evident about the much larger percentage of university graduates as well as high school graduates who lack the ability to communicate accurately and correctly either verbally or in writing. Although most conflict among individuals and groups of people is the result of faulty communication, there is little or no effort in the educational system to force students to adhere to even basic rules to enable all people to speak the same language clearly.

Serious problems often result from faulty communication in legal documents such as contracts, deeds, laws, and regulations. Utility managers devote a considerable amount of legal expertise and personal time to these problem areas, but at the same time they virtually ignore the adequacy or accuracy of the language used in policy statements, standing instructions, directives, letters, and memoranda. The tendency in present day verbal communication is to say "What I meant was," or "You know," or "You know what I mean," rather than just to say what is intended. Written communication, like verbal communication, requires adequate preparation and care to ensure that what is communicated is what the originator intends. Inasmuch as almost all water and wastewater utility management work is conducted through verbal or written communication, inadequate or inaccurate communication must be considered one of the more serious constraints facing the manager. It is essential to proper management of water and wastewater utilities for communication within the utility organization, with the governing body, with the public, with regulatory agencies, and with every other agency with whom the utility manager must associate in any way to be improved and maintained at a clear, correct, understandable level.

Succeeding chapters will deal in detail with these constraints and responsibilities of water and wastewater utility management. Opportunities for succesfully coping with these constraints will be presented to the reader not so much to provide prepared solutions to problems, but rather to stimulate the reader toward developing the thought processes necessary to solve problems as they develop and to avoid problems whenever possible.

Summary

The basic responsibility of the managers of water and wastewater utilities is to provide adequate water and wastewater service to the customers of their utilities at a reasonable price with maximum reliability based on adequate technical and fiscal planning.

The major constraints within which the water or wastewater utility manager must perform his duties and meet his responsibilities are regulatory, physical, technological, fiscal, energy, legal, personnel, political, public relations, and communications.

Organizational Constraints and Opportunities

The type and form of organization within which a water or wastewater utility manager is required to conduct the activities of the utility places unique and occasionally troublesome constraints on his management activities. It can also provide him with management opportunities and techniques which would not otherwise be available to him in other types or forms of organizations. If the utility is a part of municipal government or metropolitan government, for example, the manager may be limited in ways which might not be encountered in a special service district, a regional authority, or a private utility company. On the other hand, the municipal organization structure may allow the manager certain latitude or opportunities not available to managers of the other types of organizations. Similarly, autonomous utilities have their own advantages and disadvantages. This chapter contains organization charts and accompanying descriptions of the different types of utility management structure, as an aid to understanding the dissimilarities among them.

The constraints and opportunities which result from organizational structure can include political freedom or limitation, staff support, financial and budget support, sole responsibility for bond issues for construction, federal and state regulation demands, and the ability to apply specific technologies to specific needs. The many ways in which these various constraints and opportunities impact the water or wastewater utility manager in his decision-making process will be presented in this chapter.

An understanding of the advantages of some organizational structures over others can enable a manager to investigate and effect organizational changes for improving management capabilities for better serving the utility customers, and also to develop methods of better serving the utility customers within organizational constraints which cannot be changed.

Municipal Utilities

The common types of municipal utility organizational relationships are shown in Figures 2-1, 2-2, and 2-3. Illustrated in Figure 2-1 is a water and wastewater utility which is a division of a public works department in a municipal government under the administrative leadership of a city manager or mayor. The organization shown in Figure 2-2 is typical of a municipal government organization in which a utilities department is on the same administrative level as the public works department and other municipal departments. Again, as in Figure 2-1, the chief administrator of the municipality is a city manager or mayor. Figure 2-3 illustrates a utilities commission which may be a part of a municipal government, but which would operate independent of the city council and also would not be under the administrative control of the chief administrator of the municipal government. In this type of utility organization, an appointed or elected commission or board of directors would hire a manager or executive director who as the chief administrator of the utility would be responsible for the management of the affairs of the utility.

Utility Organized as Division of
Municipal Public Works Department

The utility organization illustrated in Figure 2-1 is a division of a municipal public works department and as such has a number of advantages over organizations which are not included in a public works department. One of the principal advantages is the availability of departmental services to which other municipal departments would not necessarily have routine access, including surveying, engineering study and design services, and equipment maintenance, where the utilities division need not bear the full financial burden of these activities. In large public works departments a personnel or purchasing division or an administrative division would probably provide these services and benefits to the utilities division, which normally would not need its own staff for these activities, while the economies from bulk purchasing, joint personnel services, and pooling of maintenance and equipment resources would be realized by the utilities division.

One of the disadvantages of the water and wastewater activity functioning as a division of a public works department is that the required staff and support personnel may not be available when the utility division needs them. At such times it may be necessary for the manager of the utility division to rely on staff personnel who are not familiar with specific problems

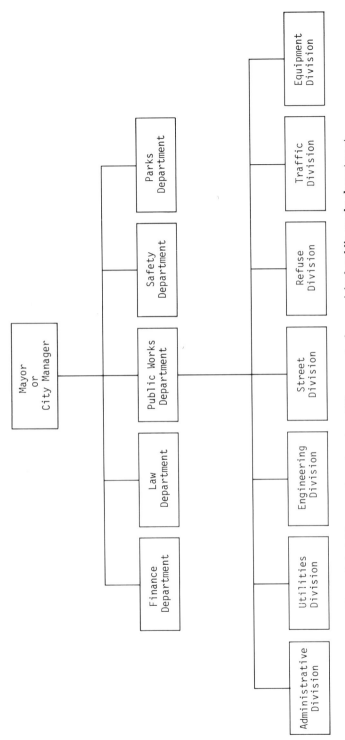

Figure 2-1 / Water and wastewater utility as part of a municipal public works department.

15

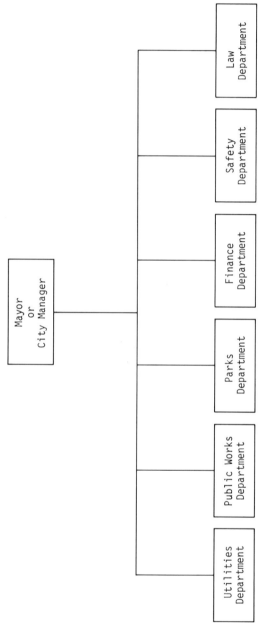

Figure 2-2 / Water and wastewater utility as a department of municipal government.

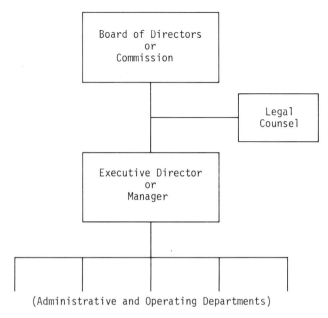

(Administrative and Operating Departments)

Figure 2-3 / Water and/or wastewater utility as a separate commission or board.

or needs of the utilities division. This can result in delays, interdivisional conflict, and inefficient use of personnel. In addition, it will usually be necessary for the utility manager to compete with various city departments and divisions of the public works department for city revenues, especially at budget preparation time or at times when a municipal bond issue is being prepared for approval by the voters, a situation which is not normally encountered when the utility activities and facilities are financed entirely by water or wastewater revenues specifically dedicated to water or wastewater activities. Various advantages and disadvantages obviously also are present with respect to a service such as equipment maintenance, which will be provided by the public works department not only to its own divisions but also to other city departments. For all such services there would obviously be competition among several departments as well as the divisions of the public works department. Finally, as long as the utility is a part of the formal city government, its management will unavoidably be subject to political pressures and the accompanying problems and constraints.

Utility Organized as a Municipal Department

Similar advantages and disadvantages will be experienced by a water or wastewater utility which is a separate department of municipal government. For example, as can be seen in Figure 2-2, it may still be necessary for the utility department manager to compete with other departmental managers for legal, personnel, purchasing, and accounting services and other assistance if the department is not large enough or is unable for some other reason to employ the appropriate personnel. The advantages of shared cost of staff services apply to the utility department as well. Further, all municipal departments are in a position to take advantage of bulk purchases, equipment pooling, and the acquisition of other resources without bearing the entire cost of the services. It should be obvious, however, that the manager of a utility department or division within a municipal government will find it necessary to plan the use of personnel or services of other departments or divisions more carefully than if he had such personnel for his exclusive use. Here again, political pressures may have a negative impact as well.

Utility Organized as a Utility Commission

The utility organization illustrated in Figure 2-3 represents a separate utility commission which would be governed by a policy-making body such as an appointed or elected commission or board of directors. It would normally receive its administrative guidance from an executive director or manager who would report directly to the utility commissioners. One of the advantages of this type of utility organization is that the commission is removed from direct political influence, in contrast to elected city councilmen. In addition, there is no competition with other municipal departments for staff assistance from, say, legal, purchasing, or accounting personnel, and the governing body of the utility commission would have only one or two areas of concern, such as water and/or wastewater, rather than the full range of responsibilities of a municipal government. Planning is less complex than it would be if it involved other city agency desires and needs, as well as direct influence from a city planning office, and the revenues required by the utility are all generated by the utility commission specifically for utility purposes, with no competition for general city funds. The lower level of political influence exerted on the commission permits the commission's decisions to be based on need and factual information, and freedom from competition for city staff assistance can mean fewer delays in many activities of the utility agency. The utility's planning, even though coordinated with appropriate planning activities of the municipal government and other agencies, will also probably not be needlessly delayed or hindered by other planning agencies.

Disadvantages of the separate utility commission include the need to provide the funds for the additional staff functions as well as for equipment maintenance and similar activities. In addition, the coordination of planning with other agencies within the same local government jurisdiction is potentially more rather than less difficult than when the utility management and the management of the other agencies is under the same legislative or administrative control.

Utility Organized as a Special Service District

An organization chart representative of a special service district such as a water district or sanitation district is illustrated in Figure 2-4. This organization would be appropriate for a separate water or wastewater utility and would serve as the organizational framework for a combined water and sanitation district if both water and wastewater services are provided by the special service district. One of the advantages of this type of organization is that it has responsibility for only one type of service, as opposed to several types of service requiring a more complex organization and additional expertise and possibly involving competing or conflicting activities. In addition, the policy-making body for the single service district would be composed of an elected board of directors which would be responsive to citizens directly; there would be minimum interference by normal municipal politics; there would be minimum interference from other municipal activities or departments; and simplified billing and accounting for a single service and a single organization would be possible. The organization is relatively free of complexity; the governing body sets policy, which is administered by an

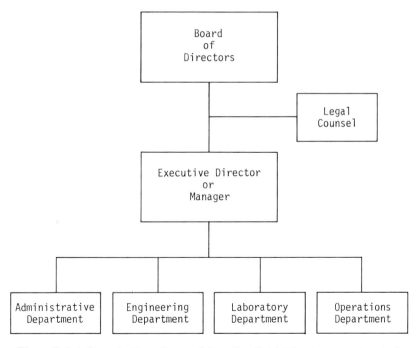

Figure 2-4 / Organization of a special service district (water or wastewater).

executive director or manager who is responsible for the activities of the various operations, administrative, and other staff divisions.

Disadvantages of this type of organization, as with the utility commission organization, include total fiscal responsibility for providing legal, accounting, personnel, purchasing, and other types of staff activities. Coordination of planning with other local government agencies can also be more difficult than if the utilities organization were part of the municipality or other type of local government. In addition, the costs of many of the utility functions will probably be higher for separate water or sanitation district operations than for a combined water and sanitation district operation, or for the operations of a utility which is part of a municipal government organization.

Organization of a Combined Water and Sanitation District

The organization illustrated in Figure 2-5 is representative of a special service district which provides both water and wastewater services. It is managed by an executive director or manager who is responsible directly to an elected board of directors of the water and sanitation district. The internal organization might include separate water and wastewater departments, as shown in Figure 2-5, or it might combine many or all of the water and wastewater operational functions into a single operational department. In the former case, the utility might be organized so as to provide for activities such as accounting, engineering, laboratory services, and construction services within each of the operating departments or the district might be so organized as to provide some of the accounting, engineering, laboratory, and construction services through departments which would provide the appropriate services to both the water and wastewater operations departments. The internal organization of water and wastewater utilities will be addressed toward the end of this chapter.

The advantages of the combined water and wastewater district over separate water and wastewater agencies are similar to those of agencies which are included in public works departments: principally, lower costs and improved coordination of planning, operations and staff support because of the shared legal, accounting, engineering, laboratory, purchasing, and other staff activities. The accompanying disadvantages include probable competition for services between water and wastewater personnel; the more complex organization required for the management of two major services rather than one (still less complex than within a public works department of a city); and the more complex system of separate water and wastewater charges, billing, and accounting as governed by Federal water pollution control legislation and other legislation or regulations which apply to only one or the other of the services.

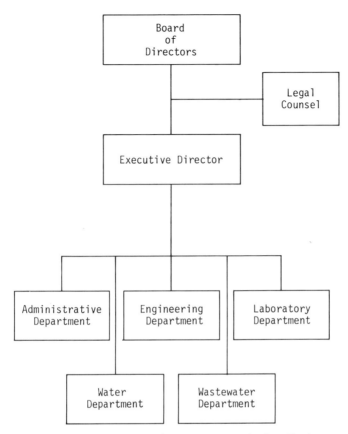

Figure 2-5 / Organization of a water and sanitation district.

Regional Water or Wastewater Authority

A regional water or wastewater authority will usually have an organization similar to that of a water and sanitation district. A regional authority provides service either directly or through governmental entities such as cities, townships, water and sanitation districts, and counties. Such a regional authority may provide direct service and direct billing to individual customers, or may provide wholesale service to several governmental entities, who in turn would provide the individual service and billing to residential, commercial, and industrial customers.

One of the intents of the Federal Water Pollution Control Act Amendments of 1972 and subsequent congressional enactments was to encourage regionalization of wastewater facilities for the purposes of improved management and operation of facilities; economy; and improved coordination of planning, construction, and operation. A regional water authority or a regional wastewater authority can often provide the mechanism for achieving the economies and management advantages envisioned by Congress.

The organization chart of a regional authority might resemble that of a water and sanitation district, as shown in Figure 2-5. The policy-making body, administrative officer, and internal organization all would be similar, except perhaps as to the number of people and positions involved.

The advantages of a regional service authority include better control of treatment within a region; lower costs of management, operations, and maintenance as well as lower charges to the individual customers because of economy of scale for the various construction, operations, management, and support activities of the authority; minimizing of contacts with regulatory agencies, with resulting reduction in conflicts and delays; the opportunity to retain the most competent personnel available for management, engineering, operations, maintenance, and laboratory because of the higher salaries the authority can pay; good planning coordination within the region; and a minimum of political influence on the activities of the water and wastewater organization.

The disadvantages of a regional authority generally are related to the greater magnitude of technical, fiscal, political, and environmental problems to be encountered by larger agencies. An example is the processing and disposal of sludge from water or wastewater treatment, where the larger volumes of water or wastewater which require treatment obviously result in larger quantities of sludge, which provide correspondingly larger problems. Other disadvantages related to size include increases in traffic problems, maintenance problems, and labor problems as the size of a facility increases.

Regionalization of water and wastewater utility service, which has already been mentioned, is worthy of further emphasis. Regardless of the problems and disadvantages inherent in the consolidation of several water or wastewater utilities into a larger single agency, the advantage to the general public of utilities regionalization can be expected to lead to increased regionalization in the field, the improved planning within a single regional agency, the economies of scale available within a single agency, and the statutory and regulatory burden placed on utility management make it essential that regionalization of water and wastewater utilities be evaluated and implemented wherever possible.

Private Water Companies

Private water companies have been established in many locations where no governmental agency was available to provide water service, and where the required service could not otherwise be obtained by individual users. Organizationally, a private water company will differ from the previously mentioned utilities in that it will typically be governed by one or more owners rather than by a publicly elected or governmentally appointed board of directors, because private water companies are often partnerships or single ownership companies. In other respects the functions of a private water company are similar to those of a special district or other public water utility.

Organizational Challenges

Governmental water and wastewater utility organizations are established in accordance with state enabling legislation or city charter, and each such organization provides its own constraints and opportunities for the manager of the utility. Political and fiscal constraints, as well as the many other constraints faced by the utility manager, obviously vary from one organization to another, and some of these contraints are more important to one manager than to others. This is also true of the many opportunities which present themselves to assist utility managers in better serving the public.

The utility manager must plan, finance, build, operate, and manage the facilities and activities of the utility in accordance with the specific constraints and opportunities of his utility organization. In those utilities in which the form of utility organization presents constraints which are serious enough to interfere with the proper conduct of business, it is the responsibility of the utility manager to provide the necessary leadership to bring about any needed changes in the organizational structure. This must be accomplished regardless of the amount of opposition to such change which may be present in the political, social, and financial environment of the utility. The regionalization of wastewater management as proposed by Congress in the Federal Water Pollution Control Act Amendments of 1972 and subsequent legislation does not mandate central operation, but does, in effect, seemingly mandate the ultimate development of regional management. Proven economies can be realized by regionalization and consolidation of water and wastewater utilities, and these benefits to the consumer far exceed the tangible costs, the required organizational effort, and other intangible costs such as the personal prestige relinquished by some influential people in special districts, small cities, and similar small organizations when the utility service is assumed by a regional agency. Decisions concerning regionalization of water or wastewater utilities should be based solely on the anticipated costs to the public and the anticipated benefits to be realized by the public.

Internal Organization of Water and Wastewater Utilities

Within a water or wastewater utility, the administrative and operational organization can both impose various constraints on and present special opportunities to managers, just as can the policy-making or legislative organization of the utility. In many cases, the relationships of a utility organization to other governmental organizations largely determines the constraints and opportunities to be encountered by the chief administrator. As an example of such opportunities, if a wastewater utility is part of a municipal public works department with a large equipment maintenance activity and a purchasing office, the wastewater utility manager will probably be able to use effectively the public works department equipment maintenance services and purchasing services. Likewise, any legal, personnel, engineering, or other staff services available from the public works department or from other departments within the city would be available to the wastewater utility manager at zero or reduced cost.

The utility organization shown in Figure 2-6 is a part of a public works department and includes both the water utility and the wastewater utility. As a part of a public works department this utilities organization could have available to it the required legal, personnel, purchasing, and accounting services as well as engineering, equipment maintenance, and construction services from other divisions of the public works department. Figure 2-6 shows the engineering, equipment maintenance, and construction sections as parts of the utilities division. In this situation the legal services as well as personnel, purchasing, and accounting services would obviously be provided by other divisions of the public works department or by appropriate city departments. In the event that the water utility is a separate organization within the city government, only the wastewater utility would be included in the public works department.

In those organizations which include both water and wastewater utilities in one division, each utility may have its separate specialized engineering and construction sections, or they may share common engineering and construction services. Likewise, the organization might allow separate planning activities in each of the utility divisions, or might share a common water resources planning activity which would provide planning activities for both utilities.

With the joint or combined utilities organization, including both water and wastewater, it is possible that a single equipment pool would provide operations and maintenance vehicles and equipment for both water and wastewater personnel. The single equipment pool also might provide equipment maintenance service, and might furnish construction equipment to both the water activity and the wastewater activity. It would be feasible and might even be economical for a separate construction section to be provided for each utility, especially if the organization is relatively large and there is the opportunity to provide personnel who are specialized in water main, valve, and hydrant installation and maintenance in the water construction section, and to also provide personnel who specialize in gravity sewer and manhole construction in the wastewater construction section.

Another combined activity for the water and wastewater sections which can result in considerable economic benefits is laboratory services. Many

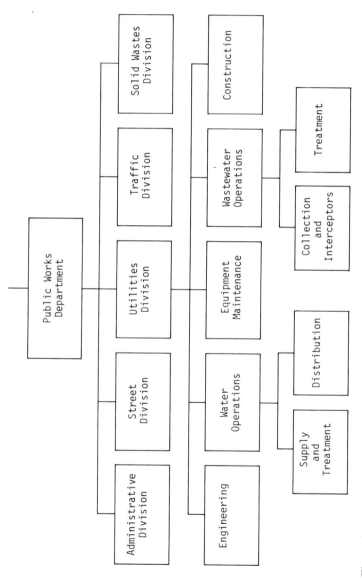

Figure 2-6 / Internal organization of a water and wastewater utility which is a division of a public works department.

25

economies can be realized by placing the water and wastewater laboratories in the same building, with some joint laboratory facilities and activities, if it is decided that a joint laboratory will not be desirable. It may, however, be desirable for each utility section to have its own laboratory, since many of the analyses performed are not common to both sections. These are a few examples of the internal organizational alternatives which must be considered in trying to find the organization which will best serve the interests of the two similar but still different utilities.

The organization illustrated in Figure 2-7 is appropriate for a utilities comission, a special water and sanitation district, a regional authority, or a private water company. In this organization the chief administrative officer would report directly to a board of directors or board of commissioners, and would be administratively responsible for all of the operational and administrative functions of the utilities organization. The water and wastewater functions are both the responsibility of the general manager under the policy control of the board of directors. For strictly water or strictly wastewater utility organizations the chart shown in Figure 2-7 still applies, except that only the wastewater or water utility is represented by the organization.

The opportunity exists in this type of utilities commission or special district organization to provide single or joint engineering services, separate or combined purchasing services, and separate or joint equipment maintenance, personnel, accounting, or other activities. It is obvious that the legal, accounting, personnel, purchasing, and probably engineering and laboratory services could be best provided for the entire organization by single departments providing the specific services to both the water and wastewater departments, as it would not be economically wise for such services to be provided individually within each of the water and wastewater departments independently.

Examination of Figure 2-7 should indicate the advantage of organizing a water and wastewater utility so that the water and wastewater operations departments can share as many staff activities as possible. Those activities which are peculiar to either the water or the wastewater department, however, will normally be provided separately in each department.

Structural Considerations

Many organizational configurations are in use for the management of all types of water and wastewater utilities. There are also numerous internal organizational structures available to the chief administrator, with all the specific constraints and management opportunities inherent in each. Advantages and disadvantages obviously exist for every external and internal organization, and it is the responsibility of the water or wastewater utility manager to maximize the benefits from the organizational advantages while minimizing the problems caused by the organizational disadvantages.

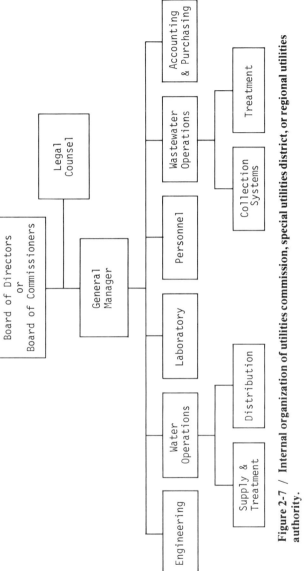

Figure 2-7 / Internal organization of utilities commission, special utilities district, or regional utilities authority.

27

Some of the features of an organizational structure are established and required by law, charter, or other legal constraint, and the utility manager is usually required to perform his management duties within these legal constraints. If the constraints are serious enough to warrant change, it then is a responsibility of the manager to provide the leadership and initiative necessary to gain the appropriate revision of the specific laws or charter.

A principle of management which is worthy of consideration is that the use of reorganization should be considered only as a last resort to solve problems of management. Nevertheless, if unreasonable constraints are placed on management because of some aspect of the manager's organizational structure, reasonable effort to provide reorganization is warranted. Such reorganization should, of course, be provided only after critical evaluation of all known effects of the reorganization has been completed.

Internal reorganization is much more common than external reorganization because the manager normally can effect the former with little or no need for obtaining approval from the utility governing body or any other authority. It is thus imperative that adequate evaluation and consideration be applied to all factors involved in an internal reorganization before such a move is taken.

Staff Positions in the Utility Organization

Within most utility organizations the use of certain staff personnel and the establishment of staff activities can provide special benefits to management and to the organization. Some of these special staff activities include public relations and public information programs and personnel, the use of administrative assistants, the incorporation of computers and data processing into the administrative and operational activities, and the use of any of a wide variety of technical specialists.

Most utility managers are content with a low level public information or public relations program and a subdued public image, but many occasions require a positive public information effort. Such occasions include bond issue campaigns which require public support; accidents or failures for which the public demands an explanation; the issuance by an individual or group of negative public statements against the utility which requires answers or explanations; and political problems which can be resolved only by gaining support of the public, the news media, or certain public political figures. Unless these and many other situations which may arise are adequately covered by public information activities, the utility can suffer serious setbacks in its programs and its ability to properly serve the public.

The effective use of administrative assistants can enable the utility manager to take advantage of opportunities to devote maximum attention to complex management considerations without losing control over routine matters within the organization. The assignment of specific duties of a routine nature to administrative assistants need not result in the loss of meaningful understanding of and communication about routine matters; it will also provide valuable experience to the administrative assistants, and will enable the manager to evaluate the personnel for the future assignment of nonroutine responsibilities. Administrative assistants with a year or more of experience with a utility organization will usually have earned the confidence of department heads and personnel of other management levels, along with sufficient knowledge of the activities of the agency to enable them to provide valuable coordination and guidance in a number of special project areas. Such administrative assistants should be considered not as trainees for future management positions, but should be considered and used as valuable staff personnel who extend the effective working range of the utility manager.

Development of computer and data processing capabilities and appropriate personnel is a necessity in the modern management of a water or wastewater utility. Modern utilities have process control, personnel, accounting, maintenance, warehousing, engineering, and laboratory activities which require the use of computers to provide better and faster information retrieval, to provide faster computations of many kinds, to permit the development of improved records and reports, and to provide more economical and efficient systems. The data processing specialist, together with the computer technicians, will work with personnel of virtually all departments to develop improved capabilities for personnel to perform their duties within each department and within the total utility agency.

In accordance with the needs of the utility organization, any of a wide range of specialists can bring a wide range of benefits to the utility manager and his agency. The areas of expertise of such specialists can include research and development, operations, maintenance, chemistry and microbiology, personnel, labor relations, management, education and training, analysis and determination of rates and charges, and many others. The modern utility manager must be on the alert to identify and investigate areas in which specialists can be used effectively.

Special Concerns of Small Utilities

Small water and wastewater utilities are normally organized along considerably different lines than large utilities. The small utilities usually have less money available to retain expertise on staff including legal, accounting, engineering, and highly skilled laboratory personnel. Specific examples of the organizational differences between large and small water and wastewater utility organizations follows.

There is sometimes a complete lack of accounting personnel and facilities within the organizations of some small wastewater utilities whose revenues are derived from ad valorem taxes collected by a county tax collector rather than from sewer use charges as a source of revenue. In this case there is no need for water meter reading, no need for billing and collection of service charges from customers, and obviously no need for accounting personnel, accounting equipment, or office space. In small organizations there is also normally no identified purchasing or personnel activity. These functions would both be administered by the manager or superintendent of the specific utility.

In those cases in which the utility organization is so small as to lack a professional engineering staff there is no opportunity for day-to-day engineering guidance for utility personnel. Consultants are retained for special projects of many types, but many technical decisions are necessarily made without the benefit of an engineering study or engineering consultation. A serious deficiency also often confronts the manager of a small water or wastewater utility in the area of operation and maintenance of facilities, because the utility is unable to compete financially with large utilities for hiring high-quality operations and maintenance personnel. This holds true also for the quantity of qualified personnel which can be retained by the large utility as compared to the small utility.

Because of the greater reliance on consultants by the small utility, the cost to the individual customer of such a utility can be expected to be greater than the comparable cost to the customer of the large utility, where much more technical expertise is available on a more economical, day-to-day basis.

The responsibility of small and large utilities is the same regardless of the circumstances. This responsibility is the provision of safe, adequate, and economical water or wastewater service to the public. Because of their more serious financial constraints, small utilities, with a smaller economic base, normally encounter greater organizational and staff problems in carrying out these responsibilities than do larger utilities.

Summary

The organizational structure of a water or wastewater utility will impose certain constraints on the utility manager in his administration of the activities of the utility, and it will also provide him with opportunities not available with some other organizational structure. It is of extreme importance that the utility manager understand these constraints and opportunities, and also that he know how to manage effectively within the constraints and with maximum advantage being taken of the opportunities.

Whenever an organizational constraint is found to be a serious deterrent to efficient management of the utility, the manager must initiate appropriate action to remove the constraint. If the constraint cannot be eliminated or revised sufficiently, the utility manager is required to develop the best method of management for overcoming the adverse effects of the constraint.

Reorganization of the utility should be considered as appropriate for improving the management and performance of the utility, but generally should be the last resort for solving problems within the utility organization.

3

Federal and State Regulation: Cause and Effect

As federal and state government have become increasingly involved in environmental matters, it has become essential that managers of water and wastewater utilities develop an understanding of applicable federal and state laws and the resulting regulations, the better to manage their agencies within the constraints imposed by those laws and regulations. It has also become increasingly important for utility managers to spend considerable time and effort to gain legislative support for needed changes in the laws and regulations. The utility manager must understand why these constraints were imposed and how they evolved, and then determine how they can be revised to provide the desired public health and environmental protection without placing an excessive and unreasonable financial burden on the private citizen.

This is not the place to examine specific laws and regulations in detail, since they will change considerably over time. It is important, however, for the water or wastewater utility manager to understand why and how the laws and regulations developed, in the hope that he can help to prevent over-regulation in the future.

Contrary to popular belief, the laws and regulations which govern public water supply systems as well as water pollution abatement activities resulted from a relatively few isolated problems in a small number of municipalities rather than from widespread problems. Publicity concerning those particular problems led to a common belief that many major drinking water problems and many major water pollution problems existed throughout the entire country.

Although the nation's legislators and regulators appear to have over-reacted to isolated situations, the cause of the overreaction was the failure of some managers and decision-makers in the water and wastewater field to solve their problems. Most major water and wastewater utilities provided adequate treatment and service with large wastewater treatment plants like that shown in Figure 3-1. Yet enough problems were found in the field to be of concern to national and state legislators as well as to a significant portion of the general public. This concern logically led to new and more restrictive laws and regulations. A lesson to be learned from this experience is that water or wastewater utility problems left unsolved will usually lead to legislative and regulatory action which will add constraints to the management of these utilities.

With the continued enactment of new water quality laws and the promulgation of implementing regulations, it is essential that knowledgeable people in the water and wastewater fields be a controlling force in future legal developments affecting their utilities. If the experts in the field do not exert every effort to guide the development of reasonable laws and regulations, it is certain that utility managers will be required to provide levels and types of treatment considerably higher than needed and beyond that level of treat-

Figure 3-1 / Large wastewater treatment plant.

ment which may be reasonable or economically achieved. The general public will then suffer because of the waste of public funds on unnecessary treatment.

Environmental Influence

As the public began to realize that the air and water in many portions of the country were of a quality either dangerous to public health or detrimental to the air or water environment, groups of people began to relay their concerns to state health departments, to national health and environmental authorities, to members of state legislatures and to the Congress.

The influence of those persons who were particularly concerned about the environment or some part of the total environment resulted in a national emphasis on protecting the environment rather than the protection of public health. Protection of the environment and protection of public health are not mutually exclusive; they can be coordinated, but some forces of society make this difficult. This is particularly true in the wastewater treatment field, where the emphasis is on protection of fishing and recreational activities rather than public health. For example, concern for ammonia toxicity to fish life in a stream receives attention, rather than concern for the presence of nitrates or nitrites, which may be a health hazard to humans. Another example is the concern for residual chlorine in a stream which may prove to be harmful to game fish, rather than for the disinfection of wastewater treatment plant effluent for the purpose of protecting public health. In addition, funds spent on construction and operation of facilities to protect fish life would perhaps be better spent on facilities to protect the public health.

The environmental influence also brought about a trend toward disregarding the balance between the cost of providing an increased level of treatment and the benefit to be derived from that additional treatment. This is seen in cases where removal requirements in the treatment of potable water supplies were far above that required for reasonable public health protection. It is also seen in cases where the level of wastewater treatment is such as to provide water quality adequate to support fishing and recreational activities, without reasonable consideration of whether the sustaining of fish life or recreation activities is possible in the water course regardless of the quality of the water. Similarly, the probable quality of the water as a result of nonpoint sources of pollution regardless of the level of treatment of wastewaters has not been given serious consideration.

Another serious disregard stemming from the environmental influence involves the adverse impacts on land, the economy, or energy sources while providing only some estimated or claimed benefit to air or water quality. These types of faulty decisions have resulted from a lack of adequate, studious consideration of the impacts on the total social and physical environment of attempts to improve only a small segment of that total environment. In many cases, there has been either an accidental or intentional distortion of the methods of evaluating needs, types, and benefits of water and wastewater treatment, resulting in decisions which were understandably erroneous and misleading.

Federal versus State Control

It has been found over the years that nationwide standards for water treatment and wastewater treatment are rarely applicable without major modification to all areas of the country. Some waterways are much more capable of providing uses such as fishing and water recreation than others, and many streams are much better able to assimilate certain concentrations of pollutants than others. Similarly, some water courses are so degraded by natural pollutants which will be present regardless of the level of wastewater treatment that additional concentrations of these pollutants from point sources have insignificant impacts on the stream. Because of the many variables involved in both water treatment and wastewater treatment in specific areas of the country, the applicability of nationwide standards to either is questionable at best.

With probably few or no exceptions, officials of state governments are much more aware of and knowledgeable concerning public health and environmental needs within their own states or in a given region of the nation than are federal officials. Hence state officials are probably better able to govern and control wastewater treatment standards and water treatment standards. However, local politics on either the state or the local level may be detrimental to local public water supply systems and water pollution abatement programs if state regulations govern water and wastewater treatment in an area with a total disregard for federal regulations and standards. There clearly must be a middle ground between total federal regulation and total state regulation wherein the public can best be served.

One of the problems which developed in the area of federal and state control involved the administration of National Pollutant Discharge Elimination System permits under the Federal Water Pollution Control Act Amendments of 1972. In many states administration of the discharge permit system was assumed by the State Water Pollution Control Agency, but the U.S. Environmental Protection Agency (USEPA) retained a veto power over the discharge permit system administration. The USEPA thus retained the power to veto meaningful negotiated discharge permits, but would not or could not provide any meaningful contribution to the negotiations between the state and local agencies to develop the optimum discharge permit, one which would adequately protect receiving waters and the total environment and yet would not burden the permittee with unnecessary or excessive costs.

Brief reviews of several representative federal acts which have had significant impacts on water and wastewater utilities are presented in the following pages.

Federal Water Pollution Control Act

In the years following World War II the American public became increasingly aware of and concerned about water pollution throughout most of the United States. The awareness and concern were conveyed to members of Congress, usually through the activities of a number of environmental groups and officials of local government. The involvement of the Congress, as might be expected, resulted in national legislation aimed at water pollution abatement throughout the country.

Various Congressional enactments led to the ultimate development in 1956 of the Federal Water Pollution Control Act, Public Law 84-660, which provided substantial federal construction grant assistance to local government for construction of wastewater facilities, and also provided for legal action to be taken against those entities who provided substantial degradation of the waterways of the nation.

The water pollution control legislation in force during the late 1950s and in the 1960s, although effective, did not stimulate some local and state governments to the level of water pollution abatement activity which was thought to be needed to reverse the degradation of some of the waterways in their areas. The failure to solve several water pollution problems which appeared to be serious, together with an overreaction by a new group of so-called "environmentalists," led to substantial pressure on Congress for new, stricter nationwide water pollution abatement legislation.

The Federal Water Pollution Control Act Amendments of 1972 (P.L. 92-500) formed the most complex piece of legislation enacted in the two centuries of the United States Congress. It contained scores of regulatory provisions, but from the date of its enactment it was improbable that it could provide nearly the benefits which the members of Congress and supporters of strong legislation had desired, expected, and promised.

The requirements and goals of these amendments included mandated secondary treatment for all public wastewater discharges not later than July 1, 1977; provision of suitable levels of treatment to provide fishable and swimmable waters by July 1, 1983; and a 1985 goal which called for the elimination of discharge of pollutants into navigable waters. The intent and goals of the amendments were commendable, but for many reasons, including bureaucratic regulatory delays, it was found to be impossible for the nation to meet the 1977 and 1983 goals in many specific instances across the country.

Public Law 92-500 included increased federal financial participation in wastewater treatment facilities construction; the implementation of a system of discharge permits under the National Pollutant Discharge Elimination System; the establishment of procedures for forcing organization of wastewater systems on a self-sustaining utility basis; a new emphasis on regionalization of wastewater management; and the development of new and improved coordinated planning on the national, state, regional, and local levels.

The construction grants program included in P.L. 92-500 provided for 75 percent of construction costs of publicly owned treatment facilities to be paid by the federal government in accordance with a detailed and complex system. The system provided for new and more detailed planning (Step I), design of facilities (Step II) and construction of facilities (Step III). Extensive review and approval of the planning phase prior to authorization of the design phase, and of the design phase prior to implementation of the construction phase, were found to delay the commencement and completion of construction of facilities. The intent of Congress was to provide financial help, generally at a level of $4–billion per year for as many as 10 years. After that time the many local governmental agencies responsible for wastewater treatment across the country would be expected to maintain a system of charges such that their revenues would be adequate to provide not only for operation and maintenance expenditures, but also the replacement of existing facilities and the construction of other new facilities as needed.

The National Pollutant Discharge Elimination System established under the P.L. 92-500 provided for a system of discharge permits which would be issued to public and private wastewater dischargers with stringent conditions contained within the discharge permits. The purpose of the discharge permit system was to enable state and federal government to stop harmful wastewater discharges rapidly and successfully, if necessary through the courts, and to provide quick and severe penalties to those dischargers who would tend to degrade the nation's waterways. Under the new discharge permit program the state and federal regulatory agencies could presumably prove violation of a legal discharge permit much more easily, much more quickly, and much more effectively than under the old enforcement system, wherein each specific legal action against the discharger required substantial proof and long litigation in the courts.

An important requirement which Congress included in the 1972 Amendments provided that public wastewater agencies must develop a system of charges to customers which would allocate to each wastewater customer its fair share of operation and maintenance costs as well as construction costs. As previously noted, Congress also required that wastewater agency income be adequate to pay for annual operation and maintenance costs and also adequate to provide for improvements, both replacements and new construction, to the wastewater system facilities. The Industrial Cost Recovery provisions of the Act required that industries which were served by publicly owned treatment works repay the federal government a proportion of the federal construction grant funds used to help construct such treatment works, to equal the proportion of the funded treatment works serving those industries.

Congress realized that the proliferation of many small wastewater agencies in many areas of the country made adequate regulation of wastewater treatment difficult, if not impossible, and thus they included throughout P.L. 92-500 provisions which encouraged regional planning and regional management of wastewater facilities. It was believed that regional management would provide more economical service to the public, would make enforcement of discharge permit conditions as well as state and federal law much easier, and in general would provide for improved management of wastewater utilities, in part by permitting support facilities beyond the reach of smaller agencies (see Figure 3-2).

Figure 3-2 / Laboratory of a major wastewater agency. Such large, well equipped laboratories are not within the financial capability of most small agencies.

The planning requirements of the 1972 Federal Water Pollution Control Act Amendments provided that each state must develop appropriate water quality standards, stream classifications, and statewide planning for water pollution abatement purposes. Each state would also be required to subdivide the state for planning purposes into regions for which a regional plan was to be developed and routinely updated so that each region would always have a current plan. Construction of new facilities or expansion or replacement of existing facilities would not be authorized before proper completion of a facility plan for the entire drainage basin within which the facility would be located, in accordance with the appropriate regional plan, which would have been approved by state and federal regulatory agencies.

Because of the complexity of P.L. 92-500, it has been extremely difficult to provide the water pollution abatement benefits desired by Congress under the provisions of the Act. Scores of implementing regulations were promulgated by the USEPA, but these regulations were so complex and voluminous that substantial revisions were necessary before the intent of the regulations and the federal law could be realized.

A valuable lesson should have been learned by officials at all levels of government in their attempts to implement the provisions of P.L. 92-500. This lesson should make it clear to everybody involved in water pollution abatement that legislation must be more thoroughly thought out than in the past, and that the implementing regulations should be required to be as brief, clear, and concise as possible. It should also be clear to federal and state legislators that water pollution abatement legislation cannot be successfully developed without maximum involvement of management personnel of operating wastewater agencies. Unreasonable legislation or unclear or voluminous regulations will only detract from and delay the implementation of the necessary activities at the local level to provide the desired benefits to the public and to the environment. The experience with P.L. 92-500 should also convince utility managers that they must be involved in the development and revision of pertinent legislation and regulations throughout the total time involved in such development and revision.

Safe Drinking Water Act

The Safe Drinking Water Act, which was enacted in 1974 and amended in 1977, answered a need of many years' standing, but just as was experienced with the Federal Water Pollution Control Act, it presented additional problems to the managers of water utilities throughout the country.

From the development of the first public water supply system to the present, water utility management has effectively eliminated virtually all waterborne diseases. The prime example of the effectiveness of public health protection by water utilities was the virtual elimination of typhoid fever, brought about by adequate water treatment and disinfection almost immediately after it was determined that typhoid fever was a waterborne disease.

For many years, the public health agencies of individual states and the nation's water utility agencies had relatively effective drinking water standards which enabled water utility managers to provide public water supply which was safe and palatable. During the 1960s a few isolated cases of detectable organic matter or chemical elements which could be suspected as health hazard agents brought about a major uproar from environmental groups, with resulting Congressional action in the form of the Safe Drinking Water Act.

The major goal of the Safe Drinking Water Act was to provide for the regulation of public water supplies so that water supplied by public agencies would be safe, and to provide for the promulgation of regulations to govern public water supplies. Some of the major specific requirements of the Safe Drinking Water Act included national drinking water regulations, primary enforcement responsibilities being assigned to states; exemptions to the requirements of the Act; regulation of underground injections; emergency powers of the Administrator of the USEPA pertaining to public water supplies; assurance of availability of chemicals needed for water treatment; research, technical assistance, information, and training of personnel; federal assistance grants for public water supply agencies; records and inspections; provision for a National Drinking Water Advisory Council; and citizens' civil action.

The national drinking water regulations were developed on the basis of the National Community Water Supply Study of 1969–70, as well as the 1975 National Organic Reconnaissance Survey by the USEPA.

The 1969–70 study indicated that although most Americans received drinking water of adequate quality, there were many Americans who did not. The presence of excessive amounts of inorganic chemicals and bacteria were found to be related to major deficiencies in state supervisory programs, inadequate monitoring by local personnel, inadequately trained operators, and antiquated treatment and distribution facilities. Specific study results revealed that more than 75 percent of surveyed facilities had not been inspected in the year prior to the study by state or county officials; 50 percent of the water treatment plant operators were not aware that they had ever been inspected by state inspectors; more than 50 percent of the physical water treatment plants provided inadequate disinfection or clarification because of inadequate capacity; the water from almost 10 percent of the treatment facilities exceeded the bacteriological standards set by the U.S. Public Health Service in 1962; and 30 percent of the treatment facilities produced water which exceeded the chemical concentrations recommended in the 1962 Standards of the U.S. Public Health Service. It is obvious from these study results that within the United States numerous deficiencies existed in water treatment operations and in water treatment facilities inspection.

The 1975 National Organic Reconnaissance Survey by the USEPA identified trace concentrations of potentially toxic organic chemicals in some surface supplies and groundwater sources; found numerous synthetic chemicals including potential carcinogens in finished waters of some municipal drinking water systems; and also demonstrated that chlorination during the treatment process could lead to the formation of chloroform, a carcinogen.

The statutory base for national drinking water standards required by the Safe Drinking Water Act includes both maximum contaminant levels and treatment regulations. The Safe Drinking Water Act also provides for public notification when the national drinking water regulations are violated, and also provides for financial resources to states for water supply regulatory programs. The resulting national drinking water regulations specified maximum contaminant levels as well as monitoring requirements for microbiological contaminants (coliform bacteria), ten inorganic chemicals, six organic chemicals (pesticides), radionuclides, and turbidity.

The primary regulations are devoted to constituents affecting the health of consumers and will be enforced either by states or by the USEPA, as appropriate, whereas the secondary regulations are concerned with aesthetic qualities and serve as guidelines only. It is assumed that the various provisions in the national drinking water regulations will be fine-tuned and revised on a continuing basis throughout the foreseeable future.

It was decided by the USEPA that they would first attempt to develop state enforcement programs for adequate monitoring and regulation of the provisions of the Safe Drinking Water Act; then continue to tighten up on sources of pollution under the provisions of the Federal Water Pollution Control Act; and finally develop a strategy and regulations concerning the protection of groundwater supplies from pollution and degradation. This step-by-step procedure was intended to result in adequate protection of the public health with a minimum of disruption to water supply management.

It is obvious that the provision of safe drinking water will require a coordinated program of minimizing the concentration and restricting the types of pollutants which can be discharged into surface waters or groundwaters, together with the development of improved treatment techniques as required.

It will be important for managers, officials, and other appropriate personnel of water supply agencies to be continually involved in the development and review of all of the regulations and guidelines in the implementation of the Safe Drinking Water Act, as well as amendments to the Safe Drinking Water Act as needed. It will be equally important for such officials to be involved in the revision and replacement of the regulations and guidelines whenever needed in the future.

National Environmental Policy Act

In 1969 the United States Congress enacted the National Environmental Policy Act, which was intended to promote a national effort to protect the environment.

The purposes of the Act, as stated in Section 2, are

To declare a national policy which will encourage productive and enjoyable harmony between man and his environment; to promote efforts which will prevent or eliminate damage to the environment and biosphere and stimulate the health and welfare of man; to enrich the understanding of the ecological systems and natural resources important to the Nation; and to establish a Council on Environmental Quality.

In the Declaration of National Environmental Policy, found in Section 101 of the Act, it is stated that it is the continuing responsibility of the Federal Government to

fulfill the responsibilities of each generation as trustee of the environment for succeeding generations; to assure for all Americans safe, healthful, productive, and aesthetically, culturally pleasing surroundings; to attain the widest range of beneficial uses of the environment without degradation, risk to health or safety, or other undesirable and unintended consequences; to preserve important historic, cultural and natural aspects of our national heritage, and maintain, wherever possible, an environment which supports diversity and variety of individual choice; to achieve a balance between population and resource use which will permit high standards of living and a wide sharing of life's amenities; and to enhance the quality of renewable resources and approach the maximum attainable recycling of depletable resources.

The main thrust of the National Environmental Policy Act was the requirement for identifying and assessing the impact on the environment of any federal programs and activities and the issuance of a formal statement of such environmental impact. The Environmental Impact Statement was required to detail the environmental impact of any proposed federal action; to indicate any adverse environmental effects which could not be avoided should the proposal be implemented; to state the alternatives to the proposed action; to indicate the relationship between the local short-term uses of man's environment and the maintenance and enhancement of long-term productivity; and to indicate any irreversible and irretrievable commitments of resources which would be involved in the proposed action should it be implemented.

The purposes of the Environmental Impact Statement requirements were commendable, understandable, and supportable, but the regulatory requirements established to implement them have proved cumbersome, confusing, and wasteful of resources. Of particular concern are the man years of personnel time and the public funds expended on the many minor details of environmental assessment at a time when these resources were needed for planning, design, operation, and maintenance of water supply, water treatment, wastewater transportation, and wastewater treatment facilities. The detailed regulations led to the lengthy delay of urgently needed water resources projects, many of which involved serious pollution of the nation's waterways, even though none of the defined adverse impacts from the proposed project could have been found to be of the consequence of the continuing pollution which resulted from the delays.

The Environmental Impact Statement process, rather than providing an improved system for abating water pollution in the United States, has thus actually served to delay, often by as long a period as several years, the construction of facilities which would have abated water pollution or which would have eliminated serious water shortages. The bureaucratic development of the Environmental Impact Statement process resulted in the expenditure of large amounts of money and the expenditure of much valuable time to meet the stated requirements of the Environmental Impact Statement process rather than to provide a brief and concise statement of environmental impact, even though the decision ultimately would be the same with the brief impact statement as with the mandated Environmental Impact Statement. Such expensive, time-consuming processes must be avoided in the future if water and wastewater utilities are to properly serve the American people. Here again, water and wastewater utility managers must take the lead in the continuing campaign for improved laws and regulations. It is imperative that in the future the construction of treatment facilities and major interceptor sewers (Figures 3-3 and 3-4) should not be delayed by bureaucratic interference.

Resource Conservation and Recovery Act of 1976

The Resource Conservation and Recovery Act of 1976 amended the Solid Waste Disposal Act, and was the result of extensive Congressional findings concerning solid waste, the relationship of solid waste to the environment and public health, and the wasting of materials.

With respect to solid waste, Congress found that technological progress and improvement in methods of manufacture, packaging, and marketing of consumer products was resulting in increasingly large amounts of material being discarded by the general public; that the disposal of solid wastes in urban areas had confronted communities with serious financial, management, intergovernmental, and technical problems; and that the problem of waste disposal had become a matter of national concern, even though the

Figure 3-3 / Construction of a wastewater treatment facility.

Figure 3-4 / Construction of a major interceptor sewer.

collection and disposal of solid wastes should continue to be primarily the function of local, regional, and state agencies.

In studying the relationship of solid waste to the environment and health, Congress found that most solid waste was being disposed of in open dumps and sanitary landfills, even though land is too valuable a national resource to be needlessly polluted by discarded materials; that disposal of solid wastes and hazardous wastes in or on the land was presenting a danger to human health and the environment; that land disposal of solid wastes was becoming hazardous to surface and underground drinking water supplies; and that alternatives to existing land disposal methods for solid wastes must be developed within the United States.

The findings of Congress with respect to materials indicated that millions of tons of recoverable material which could be used were needlessly buried each year; that methods are available to separate usable materials from solid wastes; and that the recovery and conservation of such materials could reduce the dependence of the United States on foreign resources and reduce the deficit in its balance of payments.

As a part of its energy concern, Congress found that solid waste represents a potential source of fuel which can be converted into energy; that the need exists to develop alternative energy sources for reducing dependence on natural energy resources; and that technology does exist to produce usable energy from solid waste.

It could have been logically anticipated that, based on these findings, Congress would act to rigidly control the disposal of wastes on and in the land, and to participate in the development of known and new methods for recovering waste materials and converting them into usable products and energy. This was what Congress did in enacting the Resource Conservation and Recovery Act of 1976.

The objectives of the Act were stated to include technical and financial assistance to state and local governments for the development of solid waste management plans; the providing of training grants to assist in the design and operation of solid waste disposal systems; the prohibiting of future open dumping on land and converting of existing dumps to facilities which would not be dangerous to the environment or to public health; the establishment of guidelines for solid waste collection, transport, separation, recovery, and disposal; and the regulation of treatment and disposal of hazardous wastes and similar concerns.

It was the intent of Congress to coordinate the Resource Conservation and Recovery Act with other legislation, including the Federal Water Pollution Control Act; the Safe Drinking Water Act; the Marine Protection, Resource and Sanctuary Act of 1972; the Atomic Energy Act of 1954; and similar legislation. Congress provided for the involvement of the federal government in the solid waste management field by charging the Administrator of the USEPA with the responsibility for establishing an Office of Solid Waste within the USEPA, for promulgating appropriate regulations and guidelines, and for providing assistance to states or regional agencies in developing and implementing solid waste plans and hazardous waste management programs. The Office of Solid Waste would be required to carry out the responsibilities of the federal government as set forth in the Act.

Sections 3001–3011 of the Act set forth the identification and regulation of hazardous wastes to the extent that standards could be established for the handling and disposal of hazardous wastes; the development of a permit system for treatment, storage, or disposal of hazardous wastes; and the development of state programs to control the treatment, storage, and disposal of hazardous wastes in the United States.

In addition to the regulation of solid wastes and hazardous wastes management, the Act provided for research and development activities to develop improved methods of handling and disposing of solid wastes and hazardous wastes, and provided federal financial grants for training of personnel dealing in the field of solid wastes and hazardous wastes treatment and disposal.

The implications of the Act were important to water and wastewater utilities management because of the suspected impact of solid and hazardous wastes on pollution of surface water and groundwater used for public water supplies and the inclusion of sludge from wastewater treatment facilities in the definition and consideration of solid wastes treatment and disposal. It became necessary for utility managers to become familiar with and to implement the provisions of the Act as a routine part of water supply, of sludge disposal from water treatment plants, and of the treatment, recycling, and disposal of sludge from wastewater treatment plants. The Act gave the water utility manager a new tool for protecting his water supply, and gave the wastewater utility manager new constraints within which he must process, recycle, and dispose of wastewater treatment plant sludge.

Mandated Administrative Changes and Cost Impacts

The various federal laws pertaining to water and wastewater utilities have resulted in many regulations which govern the management of virtually all the activities of water and wastewater utilities. The Safe Drinking Water Act resulted in requirements for new and higher levels of treatment for public water supplies, as well as the protection of public water sources. The National Environmental Policy Act required the preparation of detailed environmental impact statements, which in many cases caused the delay of important water and wastewater facilities construction by as long as several years. The Federal

Water Pollution Control Act construction grant conditions have resulted in delays, extra work, and increased costs for operating agencies. In addition, some permit conditions under the National Pollutant Discharge Elimination System established by the Federal Water Pollution Control Act, while providing for improved treatment and quality of effluent, also resulted in considerable additional cost to the general public.

The basic purposes of the National Environmental Policy Act were commendable, but the bureaucratic procedures for providing environmental impact statements on projects which would involve federal funds or which in some other way would involve a federal agency required detailed investigation and evaluation of factors which could not have been important enough to cause delays in urgently needed construction. As has been stated before, many examples exist of delays in construction of several years because of environmental impact assessments and environmental impact statements concentrating on secondary impacts of the construction. For example, the secondary impacts described have often involved the theoretical increase in population in an area as a result of increased water supply or improved wastewater treatment, and a corresponding theoretical degradation of air quality in such areas as a result of the increased population. Yet there is no recorded case in history where a metropolitan area has experienced growth as a result of excess water supply or excess wastewater treatment capacity. The many factors other than water supply and wastewater treatment facilities which lead to the increase in population or economic activity in an area far outweigh the importance of water supply or wastewater treatment.

The Federal Water Pollution Control Act Amendments of 1972 and the Clean Water Act of 1977 included requirements for industrial cost recovery and rigidly defined system of user charges which together delayed the receiving of federal grants by agencies which ultimately were required to delay construction because of the lack of funds. The industrial cost recovery provisions required that industries connected to municipal wastewater systems would repay to the federal government their portion of the federal grant monies which were provided to the municipality for construction purposes. This requirement applied only to industries as opposed to residential connectors and commercial connectors, and was found to incur administrative costs far out of proportion to the amount of money recovered by the federal government and local agencies. This is a good example of a provision of a law which was difficult to remove from the law even though the provision could not be supported philosophically, economically, or environmentally. The user charge requirements included in the 1972 Amendments to the Federal Water Pollution Control Act were relaxed by the 1977 Clean Water Act to permit the use of ad valorem taxes for operation and maintenance expenses under certain conditions. The purposes of the user charge provisions were to require local government to develop and maintain a healthy fiscal system for wastewater utilities and to allocate operation and maintenance costs to customers on the basis of flow and strength of wastewater received from customers. Unfortunately, the regulations promulgated under the 1972 Act contained conditions which in many situations were difficult or even impossible to implement. The total impact of these various grant conditions and other grant conditions included the delay by several years of needed wastewater facility construction, which in an inflationary economy has resulted in substantial increases in cost to the public.

The National Pollutant Discharge Elimination System provisions included in the Federal Water Pollution Control Act established a discharge permit system for wastewater agencies which improved the monitoring and enforcement capabilities of state and federal government against wastewater agencies which did not perform satisfactorily. As with many federal regulations, the discharge permit regulations permitted the imposition of discharge permit conditions which would be difficult for operating agencies to meet without the expenditure of substantial sums of money. In many cases, these conditions required substantial increases in costs with little or no identifiable benefit to the public or to the environment.

The Safe Drinking Water Act included requirements which would provide benefits in the form of improved water quality and protection of the public. But the associated regulations promulgated by the USEPA have also imposed requirements on some water utilities which would result in unnecessary expenditures which would not, in fact, significantly improve the safety or quality of the public water supply. As with other federal laws and regulations, the full impact on the public concerning health, environment, and cost was not given adequate consideration.

During the 1970s it became obvious that the authority of local government in water and wastewater matters was being eroded by the federal government. This erosion of local government authority brought with it improvements to utility management in some cases, but it also brought increased problems and costs. As the mountain of federal regulations grew, the managers of water and wastewater agencies found it increasingly difficult to use professional judgment and personal initiative in the development of methods to better treat water and wastewater, and to provide for the development of economical methods of managing, constructing, operating, and maintaining their facilities. It has become obvious that a better balance must be reached between, on the one hand, the need for federal and state government to protect the health of the general public and the environment within which we live, and on the other hand the ability of the public to pay for the levels and types of treatment required. If the requirements are based on facts and logic, the public will be willing to pay the costs. It is the responsiblity of water and wastewater utility managers to ensure that the requirements are based on facts and logic. It is also the responsibility of water and wastewater utility managers to ensure that their customers are made aware of the costs of providing specific benefits and of meeting specific regulatory requirements.

A need and obligation have fallen upon the managers of water and wastewater utilities to find ways to not only encourage, but also to participate with personnel of federal agencies in an effort to promulgate clear, concise and reasonable regulations which do not conflict with each other, and which permit and require the use of judgment and professional initiative at the local government level. For many years, local government officials have provided water and wastewater systems and facilities to protect the public health and the environment. These officials are much better qualified than federal employees to determine the treatment necessary to protect public water supplies and to provide adequate levels of treatment to wastewaters from the urban society. Local government officials thus must assume the leadership role which is their responsibility.

Need for Innovative Methods and Research

The Safe Drinking Water Act brought requirements for removal of previously unknown or previously ignored contaminants or potential health hazards in public water supplies. These additional removal requirements involved contaminants which over the years have been provided by a changing and developing industrial complex and the changing nature of the total society.

The higher levels of wastewater treatment required by the Federal Water Pollution Control Act resulted in a need for increased removal of the historical pollutants as well as a need to provide different types of treatment for nutrient removal, disinfection, and handling of increased quantities of sludge, and the requirement that certain industrial contaminants either be kept out of the wastewater stream or be removed at the wastewater treatment plant or the public water treatment facility.

Changing water and wastewater treatment needs have required new treatment and monitoring methods, which have to be provided at minimum cost. Because of the nation's energy concerns, there is also a need for recovery of resources and recycling of water and the products of wastewater treatment. These new methods could only be developed through research and development activities in all of the areas of society where the necessary expertise could be found: in the utilities field, in universities and colleges, and in industry.

Along with the need to develop new and improved methods of treatment for water and wastewater, it was necessary to provide and develop new and improved techniques for the sampling and analysis of water and wastewater so that contaminants thought harmful to the public health or environment could actually be detected, measured, and evaluated in the concentrations thought to be detrimental.

Together with management concern about the types of treatment which would be required came the realization that from the standpoint of energy conservation, resource conservation and recycling, and economics it would in many cases be advantageous to combine the treatment of wastewaters and water supplies. It appeared that the time had arrived for the public to understand that the historical treatment of upstream neighbors' wastewater as a water supply was no more desirable than the treatment of a community's own wastewater as a part of its water supply. The time had come for water utilities to use wastewater treatment plant effluent as one of its sources of raw water for treatment. How soon the technical, fiscal, legal, and educational problems of true water recycling on a large scale can be resolved remains to be seen.

New requirements for treatment methods, levels of treatment, and types of treatment bring with them the need for developing new knowledge through research as well as the training of personnel to design and operate facilities which will successfully implement the new methods. The need for new types and increased levels of treatment, together with the need for conservation of energy and the nation's resources, also requires that personnel involved in the design and operation of new treatment facilities must receive specialized training by persons competent not only in the details of the new processes but also in instructional methods and skills.

In particular, the special emphasis by the Federal Government on the use of land application of wastewater as opposed to increased levels of mechanical and chemical treatment, as well as the need for treatment of wastewater treatment plant effluent for potable water supply uses, required the development of new knowledge and capabilities for operating and maintaining the new facilities. The nationwide need for personnel who could operate and maintain the new facilities placed new responsibilities on the managers of water and wastewater utilities.

The need for additional training of personnel in the design, operation, and maintenance of the new types of treatment facilities was accompanied by a comparable need to train administrative personnel to cope with the multitude of new regulatory conditions and requirements. Public administration education and experience can provide a firm basis for effective management within the area of the normal types and quantities of government regulations, but special training of administrative personnel should be provided which can help them specifically to cope with the many new regulatory conditions which must be met by the water and wastewater utilities. In the accompanying areas of finance, public relations, and public liability, many additional responsibilities and problems also face water and wastewater utilities managers, largely because of the complex and often conflicting regulations.

This chapter has provided a brief, but perhaps critical disclosure of the additional constraints which have been placed upon water and wastewater utilities management by certain federal laws and resulting regulations. Additional laws and regulations dealing with employment, civil rights, occupational safety and health, financing, and other aspects of utility management must continually be monitored by managers of water and wastewater utilities to ensure compliance by the individual utilities. It is essential that managers of water and wastewater utilities be thoroughly familiar with the many requirements of these laws and regulations, and also that they identify areas in the laws and regulations which detract from proper operation and management of utility facilities and programs and must be changed. The utility manager must be prepared to work with the appropriate federal and state regulatory agencies, as well as with Congress and the state legislatures, to amend the various laws and regulations so that they will provide the necessary constraints for proper protection of public health and the environment and will also enhance rather than detract from the ability of utility managers to provide improved methods of protection at lower cost or at least at minimum cost.

Summary

A national awareness of environmental pollution and serious public health hazards has resulted in reactions by federal and state legislative bodies which have caused a myriad of problems for managers of water and wastewater utilities. These problems have included delays in construction of needed facilities, increased costs of construction and operation, facility design which has not been economically or technologically sound, frustration, wasted personnel time, the loss of discouraged and disenchanted employees, and many other problems.

The constraints placed on water and wastewater utility managers by federal and state laws and regulations have required the expenditure of enormous amounts of time on the part of management personnel both in familiarizing themselves with the many complex and conflicting laws and regulations and in trying to convince the legislative and regulatory bodies that most of these laws and regulations require substantial change if they are to be helpful rather than harmful to the American public.

4

Technology Constraints in Design, Operations, and Maintenance

From the time of the first effort to treat water for the purpose of furnishing a potable supply for the public, each need which was identified for improving water quality was met with one or more advancements in technology. Through the efforts of managers and operators of water supply and treatment systems, together with the advancements in processes and techniques employed by the designers of treatment facilities, improvements were made in the methods of water treatment to provide the higher quality of water required by the public. The earliest water treatment was intended to provide a water which was bacteriologically safe, was also relatively clear of turbidity and color, had reasonably good taste, and was free from objectionable odors. As people learned to expect public water supplies to be safe and reasonably palatable, they began to demand increasingly higher quality of water for reasons beyond the protection of public health. Eventually, it was found necessary to reduce the hardness of water supplied to the public, to remove iron and other troublesome constituents, and to provide better and less costly methods of removing turbidity, color, taste and odor.

A similar pattern of development occurred in the handling and treatment of urban wastewaters. When raw sewage was no longer considered acceptable in watercourses, a system of interceptor sewers and primary wastewater treatment was instituted. As urban areas developed, and the new residential, commercial, and industrial connectors contributed increasing organic loadings into the sewer systems, it was found necessary to advance to a secondary level of treatment, with contact filters and then trickling filters. In locations where even this level of treatment was not considered adequate, the use of polishing sand filters was included in the treatment train. As even higher levels of treatment were required for protection of public health and the water environment, improved methods of providing secondary treatment and eventually advanced wastewater treatment became common in many areas of the country.

The identification of the need for advanced or improved methods of water and wastewater treatment has historically led to the development and utilization of those required new treatment methods. The resultant design of new treatment processes and the construction of the new treatment facilities has always been accompanied almost simultaneously by improved procedures for operating and maintaining those more advanced treatment systems.

The imposition of new requirements for improved potable water quality and reduced waterway pollution will continue to spur the development of new treatment processes and new management, operations, and maintenance methods. It will continue to be the responsibility of the consulting engineer to use the results of research and development activities to design facilities which will include advanced processes to meet the treatment needs determined by laws, regulations, and standards for drinking water quality and water pollution abatement as well as actual public health and environmental protection needs. It will also remain the responsibility of water and wastewater utility managers to determine the levels and types of treatment needed to meet all pertinent requirements of health, environment, and economics and the demands of various political and social groups; to develop the financing needed for construction, operation, and maintenance of appropriate facilities; to schedule facilities construction at appropriate times well in advance of need; to provide the resources and train the personnel for the proper operation and maintenance of the new facilities; and to provide for the monitoring of the new treatment processes to ensure compliance with the requirements for protection of public health and the environment as well as compliance with appropriate laws and regulations.

The technology currently employed for water treatment and wastewater treatment provides adequate public health and environmental protection at hundreds of water and wastewater utility sites where the technology has been properly applied. At some locations certain environmental groups or enforcement officials may still claim that the utilities provide inadequate protection of public health or the environment, and there are some water or wastewater facilities which because of insufficient capacity, inadequate design, or improper operation or maintenance, actually do fail to meet all appropriate requirements at certain times. For the currently known waterborne hazards, however, the technology does exist for protection of public health and the environment. As new treatment standards and regulations are developed and promulgated, it will be necessary to develop new treatment technology, which will also be needed to meet needs which are found to exist in specific public water supplies and wastewater systems. Water and wastewater utility managers will be required to find and use the services of consulting engineers who have demonstrated that they can design more economical methods of meeting present day treatment needs, and also can meet the anticipated future treatment needs at a minimum increase in cost. The designs for treatment facilities which involve new treatment processes and equipment will require that water and wastewater utility managers develop adequate training and maintenance programs to provide the best possible operation and maintenance of the new facilities. The development of these programs must be accomplished well in advance of the opening of the new facilities so that adequate treatment will be achieved without delay and facilities will be adequately protected against damage or unreasonable wear or deterioration.

Process Control by Computers

As water and wastewater treatment processes have become more complex, the monitoring and control of those treatment processes have also become more complex and difficult. The use of computers to control industrial processes and processes of many other kinds has been successfully employed in a number of industries for many years. As has been true with many processes, procedures, and methods developed by and used in private industry, the successful use of computer process control in industry has led water and wastewater utility managers to investigate the feasibility of applying computer process control to their treatment facilities, transmission systems, and distribution systems. Some applications have already proved successful, and the use of computer control will probably increase considerably in the future.

Laboratory analyses for either partial or total control of the treatment process will continue to be required in many water and wastewater treatment facilities, but an increasing emphasis on instantaneous computer control of the treatment process can be expected and should be aggressively investigated for application to each specific treatment facility.

With the anticipated continuing increases in the costs of personnel, chemicals, and energy required for the operation and maintenance of water and wastewater facilities, it will be increasingly important for utility managers to improve the operation and maintenance of their facilities in every possible way to minimize expenditures. The use of computers to control the treatment process (Figure 4-1) will be a particularly important and helpful way to meet this management responsibility for cost reduction.

The actual savings in personnel costs occasioned by computer process control may not prove as great as was once thought because of the need for increased numbers and quality of personnel for maintaining and operating the process control system. Nevertheless, significant personnel savings should be realized. The savings in chemical costs and electric power costs, however,

Figure 4-1 / Control of wastewater treatment process from remote computer panel.

can be substantial. As chemical demand or electric power demand change, it will be possible with computer control to provide an instantaneous adjustment to the treatment process, so that there will be no delay in changing the process while waiting for the analysis of samples or for calculations based on sample analyses.

As in all cases of utilizing computers for specific programs or activities, it is of extreme importance for the water or wastewater utility manager to prepare for the installation of computer process control with adequate planning, design, training and employment of the necessary personnel, and retention of the appropriate consultants. The process control system must be designed for each aspect of the treatment process. A process control engineer should be employed for developing, testing, and revising the many process control programs. Maintenance personnel must be available for immediate reaction to and repair of system or equipment failures. It is also essential that backup process control equipment be furnished to perform in place of the primary equipment in case of failure.

The use of computer systems and equipment for various applications in water and wastewater utilities will be discussed in considerable detail in a later chapter.

Fitting the Treatment to the Need

An increasingly important responsibility of water and wastewater utility managers will be to operate treatment facilities according to current and future standards and regulations, however complex and conflicting, as well as to meet the real needs of public health and environmental protection in an arena of changing conditions—and to accomplish these objectives in an economical and efficient manner.

The appropriate technology for meeting water and wastewater treatment needs has heretofore been available when needed, even though the standards and regulations concerning treatment, the quality of drinking water, and the quality of wastewater effluent have continually changed. As newly discovered health hazards and environmental pollutants have been identified in water supply sources or in urban wastewaters, existing technology has been successfully applied to those new problems where practicable, or adequate new technology has been developed and utilized when needed. In the public water supply field it is inevitable that in some systems certain troublesome substances about which there has been little knowledge or little concern will be found to require removal, either because of probable health hazard impacts or because regulations or guidelines will dictate their removal. Water utility managers will find it necessary to apply proven and accepted technology to the removal attempt, and if proven technology is not adequate, muster the forces of his operations, laboratory, engineering, and research personnel to locate or develop the needed technology. If staff expertise is not available to develop the technology required for higher levels of treatment, the manager must seek out and utilize the expertise of other water utilities, consulting engineers, water utility professional associations, universities, and research institutes. When the need arises, the same approach must be taken by managers to meet the requirements of higher levels and more complex types of wastewater treatment.

The water utility manager will find that he must also select for his treatment process the technology which will best meet the treatment needs for the type of raw water available and for the types and levels of removal required for the public water supply. If the supply is from a stream or lake, it would require treatment for bacteriological safety, for color and turbidity removal, and for removal of toxic substances more than would a groundwater supply from a medium or deep well system. Treatment of a raw water supply from a deep well, on the other hand, would probably include softening and iron removal, as well as removal of other substances which enter the water from the aquifer and are considered unacceptable in potable water. The treatment might be considerably different if the raw water consists largely of wastewater treatment plant effluent. For such a source the treatment would depend on the level and type of wastewater treatment, the reliable effluent quality, and the concentration of objectionable constituents. If the wastewater treatment consists of normal secondary treatment and disinfection, treatment for potable supply will be more demanding and costly than if the wastewater treatment includes ammonia removal, dechlorination, and removal of biochemical oxygen demand and suspended solids well beyond normal secondary wastewater treatment.

The level and type of treatment to be provided to a given wastewater should depend on the use to be made of the effluent from the treatment plant. The treatment level will be set by discharge permit conditions in accordance with applicable laws and regulations, but the permit conditions will also be governed by whether the effluent is to be used for agricultural purposes, is expected to support the propagation of fish and other aquatic life, or is to be used for one or more public water supplies. As the actual uses or the anticipated reasonable uses of the effluent change, the type or level of wastewater treatment should also change if it can be determined that the total benefits to be realized from additional treatment justify the additional treatment costs. From a practical standpoint, comprehensive investigation and planning will be required to determine the highest probable use of the effluent or the receiving waters, and for designing and providing treatment to meet the requirements of that specific highest use. The decision concerning the type of treatment to be provided should be based on a complete analysis of the benefits and costs of the treatment. Even though discharge permit conditions applied to a wastewater treatment facility may be based on national or state water quality regulations and stream classifications, the wastewater utility manager has a responsibility to exert every reasonable effort to influence the decision as to level of treatment so that it will be based on actual costs and benefits to the public, the characteristics of the receiving stream, the users of the receiving stream, and the total environment.

The type of sludge treatment or processing in wastewater treatment plants is always dependent on the method or methods of final disposition of sludge. The method of processing for sludge which is to be used agriculturally as a soil conditioner is totally different than for sludge which is to be incinerated. Similarly, the processing needed for sludge to be used as a soil conditioner or fertilizer for raising crops which will be consumed by humans must normally be of a higher level than that needed for sludge which is to be used to reclaim strip mining areas in a remote area not often visited either by people or by animals in the human food chain.

For the sludge processing and disposal at a wastewater treatment plant to be economical and technologically effective, it is essential that the entire sludge processing and disposal or recycling activity be studied, planned, constructed and operated as one total process rather than as a number of individual treatment processes. Wastewater sludges usually require concentration, dewatering, transportation, and disposal or recycling. If incineration is to be used for sludge disposal, the type of sludge concentration and dewatering must prepare the sludge for ultimate incineration. Biological stabilization of such sludges obviously would not be needed. On the other hand, if the sludge is to ultimately be used for growing agricultural crops, it must be stabilized biologically, chemically, or thermally. The type of sludge dewatering must be compatible with the ultimate sludge form needed, whether powder, pellet, semisolid, or liquid. The dewatering method also will depend on the location of ultimate sludge disposal or use, whether on the treatment plant site or some considerable distance from the treatment plant. The dewatering method also must be economical as a part of the total sludge processing and disposal system. The ultimate utilization of vacuum filter sludge cake is shown in Figures 4-2 and 4-3.

There thus are numerous technological constraints which require the manager of a water utility or a wastewater utility to evaluate in detail the technology to be used in the treatment process. The source and quality of the water supply as well as concerns for specific potable water contaminants will affect the water treatment processes to be used. Additional federal or state limitations on the allowable concentrations of various substances in public water supplies will also cause the water utility manager to search out,

Figure 4-2 / Application of sludge cake to agricultural land.

Figure 4-3 / Plowing of sludge cake into soil.

evaluate, and utilize different or revised treatment technology. If required levels of treatment or contaminant removal cannot be achieved economically or satisfactorily with known technology, then new and improved technology must be found.

In the same way, the managers of wastewater utilities must evaluate and select treatment technology based on the pollutant removal needs dictated by the state water quality standards, stream classifications, the sources of specific contaminants found in the wastewater treatment plant influent, and the ultimate dispositon of the sludges which result from the wastewater treatment. Nationwide standards can be expected to impose additional requirements, which will require consideration of still other treatment technology. If available economical treatment technology is not capable of providing needed pollutant removal, the decision must be made either to apply some other technology to the treatment process or to require pretreatment of identified troublesome wastes to reduce pollutant concentrations to manageable levels in the wastewater treatment process. As is true with the water treatment plant, the treatment process of a wastewater treatment plant is established and governed by the structural facilities and equipment contained within the treatment facilities. Numerous ways are available, however, for adjusting the treatment process within the physical constraints of the facilities and also to revise the facilities to allow process changes.

It is essential for water and wastewater utility managers to continually search for and develop new and improved technology to meet the many legitimate treatment needs of the public and the many demands of regulatory agencies. Chapter 16 suggests ways of financing research and development projects which can provide new technology as needed. Chapter 16 also provides guidance to the water or wastewater utility manager for justifying needed research and development projects.

Monitoring Constraints

An important activity of both water and wastewater utilities which will require considerable attention by management is the monitoring of treatment processes and the results of treatment. Both the limitations of proven monitoring techniques and the costs of sampling and analyses should be of serious concern to utility managers. Some of the standards and regulations which determine water and wastewater treatment levels set forth the maximum permissable concentrations of specific water quality and pollutant parameters which cannot be realistically or economically identified and quantified either in the laboratory or in the field. In some cases the accuracy of a specific analysis is not adequate to meet the monitoring requirements for either the treatment process or the treated water or wastewater. Where concentration limits have been set in parts per billion and the available analytical techniques have only been accurate to parts per million, it has been necessary for more sophisticated methods of analysis to be developed. It will increasingly be necessary to develop improved monitoring and laboratory analysis techniques to enable those agencies responsible for enforcing treatment requirements to protect the public health and the environment as well as to enable those agencies responsible for water and wastewater treatment to better identify and quantify the many parameters which are of concern.

When a need arises for new or improved monitoring and analysis techniques or for new or improved treatment processes and equipment, it is essential that the required basic research, applied research, and development and demonstration work be accomplished at the earliest possible time. It is also imperative that managers of water and wastewater utilities include in their planning, for both the near and the far term, periodic reviews of sampling and analytical technology as related to probable future requirements of monitoring to control the treatment processes and to comply with standards and regulations. This type of routine review should permit the commencement of research well in advance of need for the new techniques. Such timely development will not be accomplished if the need is not recognized as a result of routine planning review.

It is also necessary to coordinate research and development activities among operating water and wastewater agencies, industry, educational institutions, and the federal and state agencies which are responsible for enforcement and research. Only with such a coordinated research and development program can future treatment and monitoring needs be met. Managers of water and wastewater utilities have an obligation to the public they serve to lead the search for new monitoring and treatment techniques, rather than to plead lack of technology and to wait for new technology to magically appear on the scene as the result of research and development efforts by others.

The Impact of Laws and Regulations on Treatment Costs

When it is found that laws, regulations, discharge permits, and other enforcement requirements contain conditions which are reasonable and economically attainable, it is the responsibility of managers of water and wastewater utilities to plan, build, operate, and maintain treatment facilities which comply with these conditions and requirements. If it is found that enforcement requirements are either unreasonably expensive or actually unattainable, then it is the responsibility of utility management to work within the legislative and regulatory systems to bring about needed changes in the requirements. It is important for utility managers to be involved on a continuing basis in the development of regulations, changes in laws and discharge permit conditions, and treatment guidelines. If this involvement is not worth the effort of the utility manager, he can only await the promulgation of new regulations or other treatment limitations and then embark on an expensive, frustrating effort to comply with them. It cannot be emphasized too strongly how important it is for management personnel in the water and wastewater fields to be continually involved in providing information in the form of written comments, statements at public hearings, and verbal or written requests to federal and state legislators and regulatory personnel on the impacts of proposed and existing laws and regulations. Such communications should form the basis of new laws and regulations whenever possible, rather than appear only in response to laws and regulations developed by people who are not responsible for economically operating and maintaining water or wastewater utilities and who thus are not qualified to enact satisfactory laws and regulations.

As was stressed in Chapter 3, many of the existing laws and regulations which seriously constrain the activities of water and wastewater utilities resulted from a lack of adequate concern, action, and forceful and meaningful input on the part of utility managers in the process of legislative and regulatory development.

The development and timely application of new treatment technology by water and wastewater utilities should accompany or even precede the imposition of new legal requirements on water and wastewater activities. It must also keep pace with new knowledge concerning individual water supply or wastewater contaminants, the effects on the public health of specific contaminants, the detrimental effects on identified segments of the environment from specific identified contaminants, and methods of identifying and quantifying the contaminants. This obviously requires that water and wastewater utility managers be imaginative, innovative, and anticipative rather than merely responsive to requirements.

It should be apparent that the various technological constraints in water and wastewater activities must be identified and overcome as they develop. To accomplish this, it is essential that the manager of a water or wastewater utility maintain a staff who collectively have the expertise necessary to study, understand, and solve the many technological problems of the particular utility. If size or financial status of the utility or some other constraint prohibits the employment of an adequate professional staff, it is mandatory that the utility manager secure the services of needed experts on a consulting basis. The types of expertise which often are required by the utility manager on a consulting basis include legal, fiscal, engineering, research, labor relations, and public information specialists. It is particularly important for the manager to have available appropriate engineering expertise in the form of consulting engineering firms who will normally perform most of the planning and design of facilities, and also the major supervision and inspection of facilities construction. It is essential that the utility manager be able to coordinate the technological expertise of his staff with the technological expertise of the consulting engineering firms who will perform the bulk of the major engineering work for the agency. The utility manager also coordinates the efforts of the several consultants who provide their legal, fiscal, public relations, and other services simultaneously on specific projects or programs.

Operations and Maintenance Physical Constraints

The operations and maintenance requirements for a specific water or wastewater utility are similar to those for other utilities, and the requirements must be satisfied if the facilities are to function properly and provide the services required by the public. The specific requirements for successful operation will include adequate personnel staffing; equipment of appropriate type and capacity in proper operating condition; adequate treatment and transmission facilities; sufficient quantities of fuel, lubricants, spare parts, and other materials; and adequate stockpiles of necessary chemicals and utility services such as electric power, natural gas, and potable water on a reliable, uninterrupted basis. If any of these operations requirements are not satisfied, it is probable that the water or wastewater utility operations will be deficient and in need of corrective action by management. Utility managers obviously must ensure that all these operations requirements are met on a continuing basis to ensure reliable, continuing service to the public.

A particularly serious type of operational constraint which is usually not easy to correct results from the design and construction of facilities which may, at a given time, not satisfactorily meet the operational needs of the utility. In such cases the decision must be made either to correct the inadequacy by supplemental construction or to adjust the process or operation to be compatible with the design or construction constraints inherent in the facility. It is essential that when these situations arise the utility manager take immediate positive action to adjust the process, revise the facilities, or embark on a construction project to make the facilities compatible with operational and process requirements. To delay decisions or positive action in such situations will only make the operational or process problems more serious.

Similar operational constraints are often encountered when equipment is no longer of adequate capacity; cannot perform satisfactorily because of inadequate speed or the lack of reverse action capability; cannot adequately handle certain types of fluids or solids; and in many other situations. It is necessary in such cases either to replace the equipment in question, or to revise the equipment (as by changing impellers on a pump), or, as in the case of incompatible facilities, to revise the operation or process to match optimum or normal performance of the existing equipment.

Many of these facilities and equipment shortcomings can be avoided by ensuring that the scope of work communicated to the consulting engineers (see Chapter 12) clearly sets forth any special requirements for matching the selection of facilities design and construction as well as equipment specifications to future levels of treatment and future processes which can be reasonably anticipated.

Even in the case of water or wastewater facilities which employ sophisticated computer process control, personnel are the most important part of the total operation of the facilities. As water and wastewater treatment requirements become more stringent and more complex, it is essential that operations personnel be provided with training and education which will enable them to understand treatment processes and provide the necessary performance. In this training it is extremely important for water and wastewater utility managers to instill the best possible positive attitudes in operations and management personnel. It is necessary to determine whether specific personnel have the aptitude to learn the new and more complex duties expected of them. Finally, it is also necessary to provide adequate funds for this continuing training of personnel. Chapter 11 provides guidance to water and wastewater utility managers in determining the numbers and qualifications of needed personnel, in the training of personnel, and in the motivation of personnel toward improvement of operation and maintenance.

A group of serious operational constraints which faces every water and wastewater utility involves the costs and availability of materials and chemicals required in the various treatment processes and for maintenance of the operational facilities. As the unit costs of chlorine, lime, ferric chloride, organic polymer, alum, or other chemicals increase, it is obvious that additional operational constraints and budget constraints confront the utility manager. The availability and cost of other chemicals and materials required for a particular treatment process can determine whether the process will be successful or can even continue. Examples of critical supplies are numerous: the media required for rapid sand filters in water treatment plants; water treatment chemicals such as lime, soda ash, alum, and activated carbon; chemicals required in the treatment and processing of wastewater sludges, including ferric chloride, lime, and polymers for sludge concentration and dewatering; chemicals which may be required for pretreatment of wastewater for enhancing the settleability of solids in primary wastewater treatment facilities; and so on. Many supply problems can be avoided by following the guidance presented in the portion of Chapter 13 which deals with purchasing procedures and contracts.

Another operational constraint which cannot be accurately predicted and which can cause major problems for the operators of water and wastewater utilities is the weather. For biological processes, it is not practicable to attempt to operate facilities with the biological activity to be on as high a level in cold weather as in warm weather. Extended warm or cold periods of weather out of season will have a significant effect on the biological treatment. A climate which involves 35 inches of precipitation annually obviously will involve operational constraints which are not to be found in a climate in which only 12 inches of precipitation are received annually. The operational procedures and treatment processes of water and wastewater treatment facilities generally will be affected and constrained by other factors, but adjustments must at times be made to respond to the impacts of climate or weather. To ignore these important factors is often to suffer sad consequences.

Because of physical operational constraints and others faced by water and wastewater utility managers it is imperative that easily understood operations manuals be developed for the use of operations personnel. Operations manuals should set forth in detail those operational procedures which have been established through successful past practice and which take advantage of the positive physical operational factors and avoid the problems which could be caused by serious constraints. The Federal Water Pollution

Control Act Amendments of 1972 and later legislation mandated the development of operations manuals as a condition of wastewater facilities construction grants from the USEPA to local wastewater utilities. Although operations manuals of some degree of adequacy were in use in some localities, the managers of many water and wastewater utilities had not adequately fulfilled their responsibility to develop manuals, but instead waited until they were forced to do so. It is essential that in the future such activities be undertaken as the result of advance planning rather than as the result of legislative action.

Satisfactory maintenance of water and wastewater facilities requires that adequate personnel, tools and equipment, spare parts and lubrication materials be available when needed for maintenance activities. An important part of providing adequate maintenance of facilities is a properly stocked and manned stockroom with a system for ordering and issuing of materials and spare parts such that maintenance personnel can effectively utilize their working hours performing maintenance work rather than waiting for materials or parts. The quality of facilities operation will be completely controlled by the quality of facilities maintenance.

The providing of employee capability to coordinate delivery of materials and tools with the assignment of appropriate personnel for maintenance activities can be achieved by following the employee selection and training advice found in Chapter 11 and by following the purchasing procedures found in Chapter 13.

A particularly important and troublesome constraint in many water and wastewater utility maintenance activities is a negative attitude among supervisory personnel or other personnel toward preventive maintenance as contrasted to corrective maintenance. In many utilities the concept of providing an effective preventive maintenance program is either unknown, not understood, or in some cases not considered to be worth the time, effort, and expense involved. The managers of water and wastewater utilities have an obligation to replace corrective maintenance as much as possible with an effective preventive maintenance program. With a preventive maintenance program which is properly developed, implemented, and routinely updated, total maintenance costs can be reduced to a minimum while at the same time equipment failure, especially breakdowns, is minimized to the point of virtual elimination. A study of the cost effectiveness of preventive maintenance will indicate to a manager responsible for maintenance of facilities and equipment that an effective preventive maintenance program is considerably cheaper overall than reliance on corrective maintenance.

An additional serious constraint in providing adequate maintenance of facilities and equipment is the lack of standby equipment or standby facilities. With adequate standby equipment provided for each critical operational function it should always be a routine procedure for maintenance personnel to shut down a piece of equipment which requires scheduled preventive maintenance and use the standby equipment in its place. The normal manner in which standby equipment should be used is to schedule the operating hours on each piece of similar equipment to be equalized as much as feasible. This should allow relatively equal wear and tear on each piece of equipment and prevent premature failure, major rehabilitation, or replacement of any of the equipment. Without appropriate standby equipment, preventive maintenance will have to be scheduled for times when it is necessary to pay overtime wages or when it will be necessary to rush the maintenance and risk inferior work, or the preventive maintenance may actually be delayed to the point where serious damage is suffered by a piece of equipment or the equipment breaks down at a critical operational time. In the latter case, serious degradation of treatment, with accompanying serious legal and political problems, can occur.

The maintenance program of a water or wastewater utility can be either enhanced or severely curtailed by the long-range planning program of the utility. If the long-range planning and capital improvement scheduling of the utility are not only adequate but imaginative, there will be suitable standby equipment and additional adequate facilities capacity to permit the scheduling of preventive maintenance on whatever facilities are involved in the treatment process without upsetting or seriously hampering the water or wastewater treatment process. The long-range planning program, based on accurate historical maintenance and operations records, should easily identify which equipment and facilities actually require standbys for preventive maintenance purposes, and which can be taken out of service for maintenance without interfering with the treatment process.

In the broad area of facilities and equipment maintenance, it is imperative that water and wastewater utility managers develop a concept of total maintenance which includes a positive attitude toward a formal preventive maintenance program. This positive attitude must be accompanied by whatever effort, time, and expenditures may be needed to develop and implement a satisfactory preventive maintenance program. Such a program must include suitable funding, adequate planning, and adequate scheduling of preventive maintenance work. With such an attitude toward preventive maintenance and the development of an adequate preventive maintenance program, it is certain that the total maintenance costs, treatment process interruptions, and related problems will decrease while the utilization of equipment and proper maintenance of facilities will increase.

Summary

Since the first water and wastewater utilities commenced operations, each new operational problem or treatment requirement resulted in the eventual development of new or improved technology to cope with the problem or requirement.

The utilization of computer systems for controlling treatment processes and other water and wastewater utility functions has permitted the utility managers to meet successfully the requirements of applying new technology to cope with most recent technological constraints and demands.

As federal and state laws and regulations have forced new treatment demands on water and wastewater utilities, these demands have been met by fitting known treatment technology to the demands or by developing, through cooperative research efforts, new technology to meet the needs caused by the demands.

New treatment and monitoring requirements have made it necessary for personnel of water and wastewater utilities to develop, in concert with others, new and improved methods of identifying and measuring concentrations of contaminants not previously of concern and also measuring concentrations at levels previously not possible with equipment normally used. As new monitoring demands are placed on water and wastewater utilities it will be necessary for the utility managers to provide equipment and personnel capable of meeting those new monitoring demands.

Inasmuch as most federal and state laws and regulations place new or more stringent requirements on water and wastewater utilities, it is essential that the managers of these utilities play a major role in the development of those laws and regulations. Without such major involvement by utility managers, major additional technological constraints will be imposed on the utilities.

The many physical constraints within which operations and maintenance activities must be performed can be minimized by adequate planning of all activities and by using the best available technology on a routine basis.

5

Total Water Resources Management

In a number of communities in the United States the management of the water utility has been combined with the management of the wastewater utility to form a single agency which has total responsibility for providing both types of service. The combining of these two utilities into one has permitted certain economies which are not available to individual water and wastewater utilities with separate administrative, planning, construction, operations, and maintenance activities.

It is not the intent in this chapter to recommend that in all communities, or even in most communities, the water and wastewater utilities should be consolidated into a single agency, or to recommend the consolidation of all water resources activities in a community into a single agency. It is hoped, however, that the considerations set forth here will enable the managers of water utilities and wastewater utilities, as well as other water resources agencies, to develop a joint water resources management attitude which can be expected to result in significant benefits.

Impact of Wastewater Treatment on Public Water Supplies

The type, difficulty, and direct costs of treatment of potable water supplies are all related directly to the source of raw water and the quality of that raw water. Raw water sources which come from deep aquifers or from remote lakes or streams are usually of relatively good quality, especially with respect to the historical water quality parameters such as bacteriological safety, taste, odor, color, turbidity, and toxic substances. Sources of raw water which emanate from deep aquifers have been subjected to natural cleaning by the various earth strata through which the water passes. Such water supply sources, of course, usually also contain elements such as high iron concentrations or hardness, depending upon the geological formations through which the water travels. Sources of raw water which emanate from remote water courses also generally have relatively high quality because the distance of the source from urban development limits the sources of pollution.

Those water supply sources which include the effluent from wastewater treatment plants or which include drainage or seepage from other urban waste sources or rural agricultural or pollutional activities often include any of a wide range of substances whose removal can require extensive additional water treatment efforts with associated costs. Special hazards also threaten the health and well-being of customers of such public water systems if the necessary special treatment is not provided for such waters. High bacteriological concentrations in wastewater treatment plant effluent obviously would require an emphasis on bacteriological safety in the water supply treatment of waters which contain such effluent. High ammonia concentrations in wastewater treatment plant effluent can ultimately result in either nitrate or nitrite problems for water supplies which include a significant amount of such effluent.

It has been found in recent years that an increasing number of industrial contaminants with suspected adverse health or environmental effects may pass through wastewater treatment and find their way into public water supply sources. Such industrial contaminants must be identified, and after being identified as harmful and found at hazardous or even questionable concentrations, must either be removed from the waste stream or be satisfactorily reduced in concentration in the wastewater treatment plant or in the water treatment plant.

A relatively new requirement of national environmental legislation and implementation regulations dictates that industries provide pretreatment of their wastes to a level such that when those wastes are released into publicly owned wastewater treatment works, the concentrations of significant and possibly harmful industrial contaminants will be removed from the waste stream or reduced to acceptable levels. Within public wastewater systems and industrial connections to those systems there are a number of places in the industrial waste stream where certain troublesome or prohibited contaminants can be removed or at least reduced in concentration. The management personnel of a specific industry with a troublesome waste could provide facilities for capturing certain contaminants before they can reach the waste stream; the industry can provide industrial pre-treatment so that all contaminants in the waste stream from the industry can be removed or significantly reduced in concentration before those waste streams are released to a publicly owned sewer system; or the contaminants from industrial wastes contained in the flow in public sewers can be reduced in concentration or removed in the treatment facilities of a wastewater utility. In the case of an industrial contaminant which is not compatible with a public water supply, it may be more economical for the specific contaminant to be removed from the raw water source in the water treatment plant rather than in a public wastewater treatment plant. The ultimate decisions of how to remove such contaminants or to reduce concentrations of those contaminants to acceptable levels should be the result of close cooperation between the appropriate agencies and industries, and should be based on maximum benefit and minimum cost to the consumer public.

The determination of the responsibility for and the location and method of removing or reducing the concentration of industrial waste contaminants should be jointly made by the managers of the industry, the public wastewater utility, and the public water utility. If the contaminant cannot be removed economically by the industry, it must be determined whether the public wastewater utility can provide the required removal in its treatment works, and if so, what the cost would be. In this case, the identified cost of the contaminant removal should be charged to the industry, and the public wastewater utility should provide the needed treatment. If the industrial process can be revised to capture the contaminant at less cost than either the industry or the public wastewater utility can remove the contaminant from the waste stream, the process revision should be the selected alternative. In the event that a water treatment process could remove the specific industrial contaminant at less cost than either the industry or the public wastewater utility, the various parties should attempt to negotiate an agreement for the treatment to be accomplished in the water treatment plant. This kind of example obviously is oversimplified, but the point remains that a joint decision in such matters usually results in appropriate treatment at minimum cost.

The joint deliberations for the determination of the responsibility for and method of industrial waste contaminant removal must include fiscal, legal, technological, and public relations considerations as well as the requirements of federal, state, and local regulatory agencies. The final decision must be the result of sincere and effective negotiation by all involved officials.

In determining the type and level of treatment for potable water supplies of raw water sources which include the effluent from wastewater treatment plants, the public health impact from the raw water sources is the most important influence on the types of water treatment processes to be considered by water utility management; consideration of the cost impacts is secondary. These public health and cost impacts are of major concern, compared to other factors, when consideration is being given to using water supply sources which either do or which in the future may contain wastewater treatment plant effluent in significant quantities as opposed to other sources of raw water which either contain no identified wastewater treatment plant effluent or contain such effluent in insignificant amounts in proportion to the stream flow. The level of treatment provided by the wastewater treatment plant and the actual realistic and reliable quality of the effluent from the wastewater treatment plant with respect to prospective water quality problems will be major factors in these determinations.

Any decision concerning the development of specific raw water sources for public water supplies requires a comparison of the costs of developing remote sources and transporting the raw water to a point of treatment against the probable higher costs of treatment of raw water sources taken from nearby streams which contain significant concentrations of pollutants from urban development, including effluent from wastewater treatment plants. This type of decision will be based on many factors, but the primary consideration normally will be the total expenditure required for the extensive transportation to a treatment plant for raw water from remote sources, plus the cost of treatment, as opposed to the total expenditure for constructing and operating the more complex treatment facilities needed for removal of contaminants found in raw water sources impacted by urban waste products, a level of treatment which presumably would not be required for remote sources of raw water.

A major concern which must be addressed in using effluent from wastewater treatment plants as a raw water source going directly or indirectly into a water treatment plant and into a potable water system involves the public acceptance of the seemingly distasteful prospect of recycling one's own waste back into his potable water system. In addition to the negative psychological reactions of the public to such a concept, several technological concerns also exist for certain health-related contaminants which conceivably could carry through the wastewater treatment plant process, and also through the water treatment plant process. These contaminants, if a legitimate concern, could result in possible health hazards to the customers of the water supply system. All of these concerns, and any other concerns, must be completely and openly investigated, possible hazards identified regardless of their severity, and action taken to protect the public water supply from the identified hazards.

Rarely does a water utility manager have the opportunity to develop his future water supply planning in such a way as to include the economical use of remote and pristine water supply sources, regardless of the size of the water utility or its geographical location. The great majority of those water utilities which receive their raw water from rivers or lakes obviously use water which has come from urban areas, including, in most instances, a certain amount of effluent from wastewater treatment plants. Even remote mountain sources of raw water are polluted to some degree by animals, plant life, hunters, fishermen, and hikers. Groundwater sources also include to varying degrees contaminants from wastewater treatment plants, industrial waste treatment plants, and solid waste disposal sites, as well as other types of rural and urban pollution. In virtually all of these cases, there is no control by the water utility manager over the quality of the water coming from these various sources. It obviously would be advantageous to the water utility manager to be involved in the decision-making process for the handling and treatment of solid and liquid wastes so as to minimize the health, environmental, sociological, and economic impacts of waste disposal activities on his water treatment process. It also would be advantageous to the wastewater utility manager to be involved in the decisions concerning the uses of water taken from a stream into which his treatment plant effluent is discharged.

With many benefits obviously to be derived both by water utility managers and wastewater utility managers from the cooperative making of decisions concerning both water treatment and wastewater treatment, it is essential that personnel from both utilities be involved in joint planning activities for those utilities and in the decision-making process of both utilities.

Need for Coordination of Water Supply and Wastewater Treatment

Wastewater treatment plant effluent historically has been discharged into streams, lakes and the oceans, and water supply historically has been derived from lakes, streams, and groundwater aquifers. With wastewater treatment plant effluent carrying various types and levels of pollutants into the waterways from which water supplies will continue to be taken, it will be increasingly important that the natural purification factors in waterways, geological strata, and elsewhere in the environment, plus normal water treatment plant processes, be augmented by more advanced water treatment processes for providing safe and palatable drinking water supplies for the public.

Up until the time of rapid, high-density urban development the natural biological activity in waterways and the cleansing action of geological strata provided adequate removal of harmful pollutants from the wastes which emanated from the nation's urban communities so that these waters were acceptable for collection and treatment for public water supply purposes. Both surface water supplies and groundwater supplies were of a quality which permitted relatively simple and inexpensive water treatment processes to furnish safe and palatable water supplies to the public.

As urban areas increased in population and size it was found that the waterways in those areas were being loaded with increased concentrations of bacteriological and toxic pollutants as well as other pollutants which made the furnishing of a palatable water to the public difficult. Industrial development in urban areas likewise provided increased flows and accompanying stronger loadings to waterways, as well as new types of pollutants previously unknown. These larger total quantities and new types of pollutants required that additional treatment processes and monitoring techniques be developed within both the water and wastewater utility fields to provide safe water treatment for public water supplies.

The increase in biological, physical, and chemical loadings to waterways from industrial, commercial, and residential wastewaters made necessary the development of more stringent requirements for wastewater treatment. The resulting complex and sophisticated types of treatment were accompanied by increases in the costs of construction, operation, and maintenance, and also generated interest and concern on the part of managers of water and wastewater utilities in the possibilities of joint planning of their systems, their processes, and their organizations. Combined billing for water and sewer service charges has been commonplace for many years, and the recognition of the benefits of the concept of coordinating water and wastewater utility planning would appear to be an expected and understandable result.

Severe water shortages have occurred during the latter part of the twentieth century throughout many parts of the United States, especially in the western states, and these water shortages brought a new emphasis on the use of wastewater treatment plant effluent for water supply sources. It was found that retaining and treating the wastewaters from a community could be considerably less costly than the acquisition, development, and transportation of new raw source waters into urban areas. It became evident that the use of wastewater treatment plant effluent as a partial or even a major source of raw water supply would result in situations in which the control of the quality of wastewater treatment plant effluent would be of great concern to the water utility manager. The need for water and wastewater utility management to work together in all areas of operations, planning, financing, and public relations soon became obvious.

The basic planning for transportation and treatment facilities is identical in most respects for water and wastewater utilities. Both have been involved in long-range planning of facilities, funds, personnel, and other resources to maintain adequate, reliable service to their customers throughout future years. The great majority of this planning has been accomplished without any significant coordination of individual utility planning efforts. There thus appears to be much waste of resources and money in separate and independent planning by both agencies. Both water and wastewater utility planning rely on population location and growth projections, estimates of per capita water consumption, estimates of commercial and industrial water consumption per unit of area, water conservation impacts, and all the other factors which impact on urban utility planning, such as economics, social concerns, politics, and land development expectations. Since so many of the activities of water and wastewater utilities are identical, it was to be expected that many utility managers would attempt to coordinate and consolidate their efforts.

The message concerning the interrelationship between water supply and wastewater treatment should be clear. Rarely are decisions concerning sources, types of treatment, and quality of the public water supply unrelated to the level of wastewater treatment, industrial waste treatment, and other urban activities upstream or uphill from the raw water source. Sound management of water and wastewater utilities requires that the planning and operations of each of the utilities must be well coordinated with the other.

Consolidation of Water and Wastewater Utilities into One Water Resources Agency

The impact of wastewater treatment plant effluent on potable water supplies varies across the country, but, as has been discussed earlier, wastewater treatment plant effluent directly or indirectly finds its way into the sources of most potable water supplies. The most obvious situations of this nature involve water treatment plant raw water sources coming from large streams or lakes into which the effluent from a wastewater treatment plant is directly discharged. Wastewater treatment plant effluent also finds its way into groundwater, the same as any other surface waters, for example, storm runoff. The use of groundwater thus requires controls and concerns similar to those required by the use of surface waters.

Wastewater treatment plant effluent along with other wastes and surface runoff from urban and rural areas will certainly be a part of the raw water sources of many public water supplies in the future. It is thus essential that the level of treatment of wastewater and its cost, along with the required treatment and associated costs for handling other waters from urban areas, be included in the consideration of the nature and cost of treatment of these waters in a water treatment plant. The removal of undesirable contaminants must be accomplished in both water and wastewater treatment, but in many cases it would be desirable for certain removals to occur specifically in one of the plants rather than in the other, regardless of the cost of that particular type of treatment. The total cost to the consumer who receives both water and sewer service should be the governing factor, not the cost of the specific water treatment or wastewater treatment. It should be apparent that the decisions which establish the levels of treatment in the water plant and the wastewater treatment plants must be jointly made by the officials of both utilities, or at least be a decision which can be imposed on both managements.

For many years water meter readings have been used as the basis for billing not only for water service, but also for wastewater service, and present indications are toward a continuation of the practice. In many communities the billing for many water and wastewater utilities has also been provided on a single combined bill. This joint billing procedure, with accompanying cost savings for the water and sewer customer, is one more example of coordination and consolidation of activities and services by the water and wastewater utilities.

The more sophisticated types of wastewater treatment, especially chemical and physical-chemical types of treatment, are similar to many of the types of treatment found in water treatment plants. As the quality of wastewater treatment plant effluent approaches the quality of drinking water, as has been required in numerous locations by discharge permit conditions, water quality standards, and the Federal Water Pollution Control Act, treatment of wastewaters to provide virtually nonpollutant quality of effluent will soon be virtually the same as that used for removal of the same pollutants in water treatment plants. The costs of such additional treatment of wastewaters can be expected to equal or exceed the incremental costs of the comparable additional treatment in water treatment plants. Without proper coordination of the treatment of water supplies and wastewater, it is probable that much duplication of treatment will occur. The public deserves careful and prudent management of the funds expended by water and wastewater utilities, and the public will certainly demand the ultimate in management, control, coordination, and even consolidation of the treatment and control of expenditures for these two important utility services.

In both types of utility it is essential that water conservation practices be encouraged and recommended by management, and for numerous good reasons adhered to by the general public. As the amount of water used within a particular urban area increases, and as the amount of wastewater discharged into the corresponding urban sewer system increases, technical and fiscal planning by management of both types of utility must provide for the expansion of treatment, collection, and distribution facilities to ensure that they will be adequate for providing the water service and sewer service to the expanding area. As construction costs as well as operations and maintenance costs continue to increase rapidly, it is especially important that water conservation programs be an integral part of the total management and operation of water utilities and wastewater utilities.

In a similar vein, as water and wastewater treatment become more complex and more costly, it is increasingly important that the total quantities of water and wastewater which are provided treatment should be reduced to an absolute minimum. This might be achieved by bypassing those portions of water treatment which would not be required for providing water of less than potable quality where the needed quality is not so high, and thus avoiding the cost of a portion of the normal water treatment for potable water. Any quantities of these lower-quality waters which would not be discharged into the sewer system obviously would not require treatment in the wastewater treatment plant, and this would result in wastewater treatment cost savings. It certainly should be apparent that identical water conservation programs would be sponsored for identical reasons by both water and wastewater utilities. This, then, is an area in which joint or common management could be beneficial and effective.

In reviewing the technical and fiscal planning required for water utilities and wastewater utilities, it is apparent that virtually all the factors included in such planning are identical or nearly so for the two utilities. As a basis for technical and fiscal planning each utility manager must consider the growth of the urban concentration in terms of area; population increase; population location; -business and industrial location, type, and size; and the probable impacts of these factors on water supply needs and wastewater treatment needs. Comparable data pertaining to per capita water consumption and water consumption per unit of area, as well as assumed unit rates of water consumption for industrial production, would also be virtually the same for the water utility and the wastewater utility.

Virtually all other aspects of utility management would be similar for water and wastewater utilities: derivation of revenues from service charges, taxes, and other common sources; public relations and information programs; political impacts on the utility; probable federal and state regulatory impacts; and so on.

It should be obvious that there are many common grounds on which joint or common management of water and wastewater utilities could be effective and could provide increased benefits to the utilities customers. In the interest of providing maximum benefit to the public, these many areas of common effort and concern should be addressed jointly by utility managers.

Consideration and evaluation of the desirable aspects of sharing of administrative resources, maintenance facilities and personnel, vehicular maintenance facilities, billing and accounting resources, legal personnel, personnel administration activities, and purchasing activities should result in an increased awareness of and interest in the consolidation of water and wastewater utilities into common water resources agencies or utilities. It has been the experience in many cases over past years, as has been discussed, that many economies are realized from consolidation. There is thus every reason to believe that an increasing number of water and wastewater utilities will merge in the future, or will at a minimum improve coordination of their many administrative and operational functions. For this type of consolidation to be of benefit to the general public, it will be necessary for the managers of individual water and wastewater agencies to plan and act with the public in mind, rather than with an eye to individual or organizational advantage. There will be an understandable reluctance to promote or even support a consolidation which could result in a loss of prestige, a loss of political or economic power, or the possible loss of a secure position of employment. However, the effects on public health, the environment and the total society must be considered above anything of a personal nature in the evaluation of a utility consolidation.

Inclusion of Drainage and Flood Control into a Water Resources Agency

In the consideration and evaluation of a possible consolidation of water and wastewater utilities into single water resources agencies, it would be reasonable to evaluate the possible benefits or impacts of the inclusion of drainage and flood control activities in the water and wastewater activities of the consolidated agency.

At the two hydrologic extremes in an urban area are flooding conditions, with serious inconvenience, property damage, and loss of life at one end, and at the other, drought conditions or water shortages, with serious economic, social, and health problems. Not only do the excess waters which flow through and around an urban area during periods of flooding often cause serious damage to public and private property and constitute a hazard to life and well-being but they are also lost forever as far as any possible beneficial use to the urban area and the surrounding rural area is concerned. During periods of drought or severe water shortage an urban area could benefit considerably from the retention of flood waters for use during the dry periods. Facilities whose primary purpose would be the retention of flood waters could also serve recreational purposes, for example, fishing and water sports, in addition to the flood control and water supply purposes.

It should be apparent that urban leaders can and should be involved in at least considering the development of a water resources agency which would include water supply activities, wastewater treatment activities, and flood control and drainage activities. A water resources agency of this type would be responsible for all the activities involved in the coordination of water supply, drainage and flood control, and wastewater collection and treatment. The retention of runoff waters would be coordinated with water supply needs during those times when the public water supply source or sources would be short of raw water. The retention ponds could also provide water pollution control benefits in retarding erosion, in providing some biological benefit, and in providing storage of effluent from wastewater treatment plants.

Many joint uses of common facilities are available to water resources agencies as discussed here. There are many opportunities for coordination of activities of mutual responsibility in the areas of drainage and flood control, water supply, and water pollution abatement. The economies to be realized in urban areas by combining these three services in a single water resources agency should be apparent. Here again, the replacement of two or three administrative staffs with one, the replacement of two or three maintenance activities with one, the replacement of two or three purchasing and personnel administration activities with one all should be adequate incentive to the management of the three agencies to consider joint planning and coordination or consolidation of some or all of their activities into a common water resources agency.

The use of retention and detention ponds for recreational purposes has been mentioned as a local government activity which should be considered in connection with urban drainage and flood control as well as water and wastewater utility planning. Flood control and drainage areas generally should be zoned for the prohibition of most types of building within the flood hazard areas. The designation of such areas for use as retention ponds or detention ponds has often resulted in the use or at least consideration of these areas for recreation, a use almost always acceptable to the public. This use and similar types of use should be encouraged whenever possible for flood hazard areas.

Joint Processing of Water and Wastewater Sludge and Solid Wastes

An important area in which joint planning and also probably the coordination of operations should be considered lies in the broad activities of solid waste processing and disposal and the processing and disposal of sludges from wastewater and water treatment plants. The joint processing of sewage sludge and solid wastes has been the subject of at least some small-scale research, which has indicated that such activities promise to be successful and economical. Provisions of the Resource Conservation and Recovery Act make it clear that the joint processing of sewage sludge and solid wastes is recognized as a feasible activity and is encouraged by the federal government. This is one area of endeavor in which the officials responsible for treatment of wastewaters and the processing and disposal of solid wastes can provide substantial benefits to the citizens they serve and in fact to the entire nation by turning the costly disposal of waste products into the development of valuable national resources and financial benefit.

Inasmuch as the processing and ultimate disposal of solid wastes, water treatment plant sludge, and wastewater treatment plant sludge from urban areas all present seemingly insurmountable problems on a nationwide basis, it is not only desirable but essential that joint planning be developed and maintained on a continuing basis among the three agencies involved, namely, solid wastes collection and disposal activities, water utilities, and wastewater utilities.

The planning and development of programs and the contruction of facilities for resource recovery from the solid wastes and the sewage sludge from many of the nation's urban areas will certainly grow over the near future. This type of resource recovery is already proceeding in many sections of the country. How water treatment plant sludges will fit into the picture remains to be seen. The neutralization of hazardous wastes, such as highly acidic wastes with a lime sludge, may be required. Regardless of how it may be accomplished, it is obvious that thoughtful consideration for and joint planning of water treatment plant sludge utilization and disposal along with all other types of urban wastes should be a high priority item with managers of the various activities. This is particularly important for managers of water and wastewater utilities.

Need for Continuing Efforts at Utility Consolidation

Many economies can be realized from the consolidation of water and wastewater utility activities into a common agency. This basic fact should be apparent to utility managers. The advantages of consolidation or coordinated management should result in a considerable movement toward consolidation among water and wastewater utilities in the future. In many cases local political obstacles and public attitudes will prevent total legal consolidation of the utilities into a single agency, but many benefits to the public can be realized from cooperation and continuing communication between utility managements. At a minimum there should be common vehicle and equipment maintenance, common purchasing activity, common personnel administration activity, common planning and accounting, or joint use of other resources, at significant savings to the general public. The degree to which these activities can be shared will vary from community to community, but the benefits will always accrue to the consumer.

It is essential that both the technical and fiscal planning of water and wastewater utilities be coordinated so that both utilities offer and provide service in the same area for the same residential, commercial, and industrial customers, with a similar or even identical revenue-producing base. This will be especially important as more complex and more expensive types of treatment may be required of either utility. Joint consideration by the utility managers of the total cost to the consumer must be an important factor in determining the types and levels of treatment which should be employed in their respective facilities. An increase in the cost of either water or wastewater treatment must not deter a decision if such increase results in a decrease in the total cost to the public of combined water and wastewater service. One such situation will be found in future years in communities where wastewater treatment plant effluent will be used as a source of raw water for the local public water system. If the cost of treating wastewaters which are to be subsequently provided treatment in a water treatment plant can be reduced by more than the water treatment cost increase, the total cost to the public of water treatment plus wastewater treatment will be the determining factor in decision-making, not the cost of either water or wastewater treatment.

The national need for serious consideration of total reuse of wastewater treatment plant effluent in some urban areas will require the location of water and wastewater treatment facilities in close proximity to each other, much more so than in the past. Contiguous siting of the two treatment facilities should ultimately result in physical and organizational merger of the separate water and wastewater utilities into a single water resources agency. This agency will be responsible for water supply, conservation, and reclamation, and the total water resources nature of the agency will provide substantial additional benefits to all the people in the urban area.

Summary

Most communities have separate agencies to provide water and wastewater service. In other communities where both services are provided by a single water and wastewater utility, many significant advantages have been realized by the utility managers and by the consuming public.

The health, environmental, political, and fiscal impacts of wastewater treatment plant effluent on downstream water supply agencies should encourage close coordination in the planning and operation of these agencies, if not actual consolidation of these and other similar agencies. As shortages in energy, water supply sources and other national resources become more acute, the coordination of water supply and wastewater treatment will become increasingly important. The many factors which should encourage the coordinated efforts of officials of water and wastewater utilities will eventually result in many consolidations of utilities into single water resources agencies, which may be responsible for drainage and flood control activities as well.

Managers of water and wastewater utilities have a responsibility to utilize the resources at their disposal to bring about maximum cooperation and coordination between the activities of their utility organizations and to effect the consolidation of such utilities when and where feasible and in the best interest of the citizens they serve.

Planning, Programming, and Budgeting

In 1961, with a view toward improving its planning, fiscal control, and program evaluation, the United States Department of Defense (DOD) implemented a new coordinated planning and financial control system which was expected to revolutionize the entire DOD methodology for long-range planning, for the programming of resources and finances, for the preparation and administration of budgets, and for the continuing evaluation of the many DOD programs and projects. This new system was called the Planning–Programming–Budgeting System (PPBS), and was intended to provide for the scheduling and furnishing of resources within the department on the basis of project needs and program needs regardless of jurisdictional considerations. The new system was also intended to include the full spectrum of technological and fiscal planning, programming, and budgeting of resources of all kinds, as well as thorough evaluation of objectives, programs, and projects. The evaluation would be followed by revisions where warranted.

As the result of a certain degree of success with PPBS in the Federal Government, attempts were made to adapt the system to local government. PPBS has not been successfully adapted in its entirety to local government and utilities management, but the concepts and procedures developed have been found to be of immense value when applied to the planning, programming, budgeting, and reviewing activities of water and wastewater utilities. Without exception, the application of the basic concepts of PPBS to water and wastewater utility planning and budgeting activities will result in significant benefits.

The Need for Coordination of Planning, Programming, and Budgeting

Those water and wastewater utility managers who are not familiar with and have not developed a coordinated planning, programming, and budgeting system should investigate the feasibility of establishing such a system. They should develop knowledge and experience in the coordination of long-range and short-range technical and fiscal planning, in the consideration of multi-year budget implications of such planning, in the development of budget programming, and in the methodology for evaluation of the effectiveness of programs and activities, as well as in the development of methodology for the analysis and revision of the programs, projects and objectives of their utilities.

The development and use of a coordinated planning, programming, and budgeting system is the most effective way to allocate and control resources to carry out the various programs and projects of a water or wastewater utility. This is true for most governmental and private agencies, but it is especially true for organizations such as water and wastewater utilities whose managers must plan facilities many years in advance, must analyze revenues and estimated resource requirements for several years into the future, must provide for budget preparation on an annual or biennial basis, and must control expenditures for numerous activities during each budget year.

For many years it was common practice for water and wastewater utility managers to prepare annual budgets on the basis of the history of past expenditures for the operation and maintenance of facilities and the performance of the many utility activities, to which were added the estimated costs of specific projects of construction or maintenance and replacement which appeared essential for the coming year. The use of a detailed analysis of actual resources required for specific operations and maintenance activities was not common in budget preparation. It was much easier to use past annual expenditure information together with generalized information concerning cost of living trends and the funds available for the management of the utility to form the basis for annual budgets. Similarly, much of the planning of future facilities too often commenced at the time that a crisis was identified and the construction of additional facilities was considered essential. The routine planning effort which includes the detailed analysis and use of population trends, commercial and industrial growth information, data pertaining to water consumption per capita and per unit of production, and revisions to various operations and activities was not widely understood or used in water and wastewater utilities.

The purpose of a coordinated planning, programming, and budgeting system is to ensure that facilities will be designed and constructed in advance of need, and that the resources of all kinds which are needed for operation and maintenance of the facilities will be available to the utility manager when they are needed. This requires that funds be made available for preliminary planning, design, and construction of the water or wastewater facilities in accordance with a schedule based on sound planning data and procedures. It also requires that the needed revenues, personnel, equipment, and other resources which are necessary for sound operation and maintenance of all facilities be scheduled and budgeted for the years when they will be needed. In addition, it is essential that the utility manager develop and use a system of expenditure reporting and control which will ensure that the expenditure of resources not only in each budget year, but on a continuing basis, will be in accordance with the best interests of the water or wastewater utility and the general public.

Long-Range Planning

Modern management of water and wastewater utilities cannot be effective in the absence of adequate long-range planning, nor can any utility manager meet his responsibilities to the customers of the utility unless he routinely uses the results of an effective and continuing long-range planning effort to determine the resource needs of the utility. Without such a planning effort it is impossible for a water or wastewater utility to have the appropriate facilities constructed and in operation when needed, or to have the necessary revenues on hand to finance the operations, maintenance, and other requirements of utility activities in each successive year.

Any long-range planning which is to be effective must be based on reliable population projections, both as to amounts and location; on records and predictions of economic conditions of the past, present, and future; on a thorough understanding of the political attitudes and their implications within the urban area; on knowledge of probable future commercial, industrial, and residential development in the urban area; and on adequate information concerning the host of other factors which determine what type of urban environment can be expected during the planning period, ten to fifty years in the future.

The long-range planning period selected can vary considerably as appropriate to the planned activity, or activities. For most water and wastewater utility long-range planning a period ranging from twenty to fifty years should be utilized. For long-range water supply development purposes, particularly the construction of facilities such as dams, lake or reservoir intakes, and transmission mains, the planning period probably would be fifty years, while long-range water treatment plant expansion planning periods may be ten to twenty years. Using a modular approach to treatment facilities expansion, it is feasible to use ten-year planning period increments for expanding treatment facilities, whereas the planning for pump stations may employ twenty-year periods or longer for the pump station structure, with ten- or fifteen-year planning periods used for equipment additions or alterations. Similarly, the long-range planning period for interceptor sewers, the various parts of a sewer collection system, and sludge disposal systems would normally be about fifty years, and the planning period for wastewater treatment facilities might be ten to twenty years. Major facilities which would be constructed underground and remain buried for many years, where the construction of additional capacity is possible through most major urban areas only at considerable cost with parallel construction, as in the case of interceptor sewers, should be designed in terms of the needed capacity for the longer, fifty-year period. Treatment facilities which can be expanded by the addition of one or more treatment unit modules in an orderly, pre-planned manner can be designed for a shorter span, probably the economic ten-year period.

Reliable and effective long-range planning can be accomplished for either water utilities or wastewater utilities only with full knowledge of and adequate consideration being given to the known treatment levels which will be required during the planning period, the special treatment requirements which may be faced, and any reasonable requirements for probable changes in treatment levels which may be encountered by the utility during the planning period.

In addition to the need for projecting changes in population quantities and locations for estimating water consumption and wastewater contribution quantities, it is necessary as a part of a long-range planning program to analyze and project the probable changes in per capita water consumption and per capita wastewater quantity and pollutional loading contribution trends. Such projections must take into account the long-range historical trends; the recent short-term trends which may have resulted from excessive groundwater, unusually light or heavy precipitation, or significant water shortage situations; and the probable impacts of water conservation programs, ever increasing energy conservation concerns and reactions, and recent changes in economic conditions. Similar projections must be made for commercial and industrial water supply demands and wastewater contributions to the public sewer system. Such projections must be based on known as well as anticipated new commercial and industrial connections, anticipated changes in individual commercial or industrial processes or uses, and anticipated water conservation or wastewater pretreatment practices.

As is true for the other types of urban planning, it is essential that the long-range planning of water and wastewater utilities be coordinated with planning for transportation, land use, schools, other utilities, and land development in the urban area. Managers and planning personnel of water and wastewater utilities can anticipate that some type of regional planning agency will usually provide probable land use patterns for the future, transportation and utility planning on a regional basis, regional population and land development trends and projections, as well as coordination services with respect to the planning activities of other governmental agencies in the region.

The long-range planning effort of water and wastewater utilities should result in the periodic preparation of a long-range facilities plan which will set forth the types of facilities for the utilities which will have to be constructed within specific time frames, along with the estimated costs of such construction. This effort will require the determination of the capacities of various types of facilities such as raw water intake structures, dams, transmission mains, treatment facilities, storage facilities, distribution systems, and various distribution system appurtenances for water utilities, as well as collection sewers, interceptor sewers, pumping facilities, treatment plant facilities, and sludge processing and disposal or recycling facilities in the planning of wastewater treatment utilities. When the necessary inventory of facilities, projection of needs and estimates of costs and anticipated revenues have been completed, a schedule of facilities to be constructed within specific time periods can then be prepared. The schedule of facilities which will be required and which are expected to be constructed should chronologically list each facility to be constructed; the year for completion of planning, design, and construction; the estimated cost of each project; and any other helpful information such as the possible sources of funds for the projects. Such a long-range facilities plan should form the framework upon which all facilities planning, programming, and budgeting of the water or wastewater utility will be based. As the long-range facilities plan is periodically revised, the programming and budgeting activities will be revised accordingly to ensure adequate funding for the operation and maintenance of facilities and appropriate administrative functions, as well as for the design and construction of the facilities.

Figure 6-1 is a simplified example of a long-range facilities plan for a hypothetical wastewater utility. The example includes treatment facilities, interceptor sewers, and collector sewers similar to the types of facilities construction required by most wastewater utilities. Detailed fiscal planning for the construction project will require consideration of the estimated cost of construction at the middle of the estimated construction period.

One of the tasks which must be included in the long-range planning effort is the investigation of probable technology which will be available for water treatment or wastewater treatment within specific time frames, particularly within the planning period, and determination by management of whether the available technology can be successfully applied to future treatment needs to result in the desired or mandated level of potable water quality or wastewater treatment plant effluent quality. If the technology expected to be available at specific times is determined to be either inadequate to meet

Pine City Wastewater Utility
Long-Range Facilities Plan 1980–2010

Project Description	Estimated Cost	Construction Period
Treatment Facilities		
T16. Computer process control	$ 5,000,000	1984–1986
T17. Secondary capacity (20 mgd)	30,000,000	1992–1995
T18. Anaerobic digester capacity	4,000,000	1992–1995
T19. Headworks improvements	6,000,000	1998–2002
Interceptor Sewers		
I32. F St. interceptor	2,000,000	1982–1984
I33. 3rd Ave. interceptor	4,000,000	1982–1984
I34. 18th Ave. interceptor	1,500,000	1983–1985
I35. J. St. lift station	3,500,000	1985–1988
I36. 32nd Ave. interceptor	6,500,000	1986–1988
I37. Trout River crossing	4,000,000	1987–1990
Collector Sewers		
C76. A St. sewer (86th–90th)	800,000	1980–1982
C77. B St. sewer (14th–20th)	500,000	1980–1982
C78. 88th Ave. sewer (D–L)	1,200,000	1981–1983
C79. A St. sewer (90th–97th)	1,000,000	1982–1984
C80. 76th Ave. sewer (F–G)	600,000	1982–1984
C81. 43rd Ave. sewer (G–P)	1,300,000	1984–1986
C82. 57th Ave. sewer (C–X)	4,500,000	1985–1987
C83. G St. sewer (43rd–66th)	3,500,000	1986–1988

Figure 6-1

treatment requirements or questionable as to adequacy, and if investigation reveals a lack of suitable research and development activity elsewhere, then the manager of a water or wastewater utility should ensure that the necessary research and development efforts are funded locally or in some other manner so that appropriate technology will be available when needed.

A vital part of the long-range planning effort is the continuing review by the managers of water and wastewater utilities of the anticipated revenues and the need for changes in revenue amounts and sources for operation and maintenance requirements as well as for the capital expenditures of the utility. It is necessary to determine which, if any, additional sources of revenue may be available for use only for the capital expenditures or for operations and maintenance expenditures and which revenues can be expected to fund both capital needs and operations and maintenance needs. It is the responsibility of water and wastewater utility managers to continue this ongoing investigation so that on an annual basis the required revenues will be available through service charges, plant investment fees or water and sewer connection fees, construction grants-in-aid from federal or state government, special assessments, and other sources. If water and wastewater utilities are to provide adequate service to their customers it is imperative that the utility manager schedule the reliable receipt of revenues to meet the expenditure needs of the utility on an annual basis, rather than permit the revenues to determine the level of service which can be provided.

Multi-Year Capital Improvements Program

The long-range facilities plan should provide a sound basis for short-term technical and fiscal planning for the construction of facilities for water and wastewater utilities. The long-range facilities plan identifies the specific types of facilities for which a need is projected, indicates when the facilities will be needed, and states the estimated costs of facilities. The multi-year capital improvements program provides a short-term (normally four to six years) scheduling of construction projects within a priority system based on several factors. The capital improvements program identifies which construction projects are scheduled for each specific year within the program period, the estimated cost of each project, the projected expenditures for each project for each year in the program period, and the sources of funds to finance the construction projects. Although the sources of construction funds may be limited for water and wastewater utilities, the sources should be identified, whether they are user charges, revenue bonds, general obligation bonds, federal or state construction grants, tap or connection fees, special assessments, or any other source from which funds can reasonably be expected.

The priorities of specific construction projects will be based not only on the capacities of facilities and their capability of providing the amount of treatment or other service needed, but also will be determined on the basis of the treatment level required, the physical and structural condition of facilities, the regulations and standards controlling levels of treatment and numerous other factors. Some of the other factors to be used in determining the priority of various construction projects include scheduling and coordination difficulties both within the construction program of the utility and with other agencies, political and public relations concerns, the amount of funds expected to be available in each year of the program period, and delays which may be caused by any of numerous governmental agencies or by any other situation which may arise.

After priorities have been established for facilities construction the schedule of capital improvements should be developed in accordance with those priorities. The schedule should provide information which shows a description of the facilities improvement, the years in which the specific work will be designed and constructed, the funds scheduled for expenditure in each year during the construction period for each specific project, the total costs of the specific project, and the sources of funds for financing each of the individual construction projects. The form of the capital improvements program is not important as long as it includes the information needed. It is important, however, that the capital improvements program be formally adopted on an annual basis as a document which will guide the construction program of the utility.

The capital improvements program shown in Figure 6-2, although over-simplified, illustrates the types of information which should be included in a capital improvements program and also illustrates its relationship to a long-range facilities plan (Figure 6-1). Since the capital improvements program example is for the years 1980–1984, only the projects which are scheduled for planning, design, or construction activity in those years are taken from the long-range facilities plan and included in the capital improvements program. Note that projects I29, I30, and C74 are not shown on the long-range facilities plan for the period 1980–2010 because these projects are funded and under construction. Some utility managers may wish to include such funded and under-construction projects on the long-range-facilities plan along with the "future" projects.

In the development and preparation of a formalized multi-year capital improvements program, it is necessary to coordinate the actual needs for specific construction occasioned by a need for additional capacity or by existing substandard physical condition, the allocation and availability of construction funds among the various construction projects, the scheduling of other projects which for operational reasons cannot be accomplished simultaneously within the same time period, the construction projects of other agencies which could interfere with specific scheduled utility projects during construction, politics and public information factors, and anticipated operations and maintenance cost impacts. When the planning, fiscal, and management personnel have coordinated the long-range facilities plan with the numerous influencing factors, including those just mentioned, an effective four-, five-, or six-year capital improvements program can then be developed. This capital improvements program can then serve as the official construction plan of the water or wastewater utility, and can provide the base from which the multi-year budget programming effort can proceed and which will logically lead to preparation of the annual or biennial budget of the agency.

The capital improvements program provides for the expenditure of funds which are earmarked for construction only, and does not set forth the funds required for the operation and maintenance of the facilities constructed. The capital improvements program preparation process should therefore require justification of each specific construction project not only on the basis of need and cost of construction, but also on the basis of the additional future

Pine City Wastewater Utility
Capital Improvements Program (1980–1984)

Code	Description	Location	Total Cost ($1,000)	Cost in 5 Year Period ($1000)	Cost in Year ($1,000)					Source of Funds
					1980	1981	1982	1983	1984	
T16	Comp. proc. cont.	Jones WWTP	$5,000	$1,200	—	—	—	400	800	Revenue bonds
I29	G. St. int.	2nd–14th	2,500	1,500	800	700	—	—	—	Revenue bonds
I30	8th Ave. int.	C–J St.	3,000	2,200	600	1,000	600	—	—	Revenue bonds
I32	F St. int.	14th–26th	2,000	2,000	—	—	200	1,200	600	Rev. bonds 80%, priv. funds 20%
I33	3rd Ave. int.	L–V	4,000	4,000	—	600	1,500	1,700	200	Revenue bonds
I34	18th Ave. int.	B–H	1,500	1,200	—	—	150	350	700	Rev. bonds 60%, priv. funds 40%
I35	J St. lift sta.	James & 22nd	8,000	800	—	—	—	—	800	Revenue bonds
C74	77th Ave. sewer	R–W	600	400	400	—	—	—	—	Special assessment
C76	A St. sewer	86th–90th	800	800	150	400	250	—	—	Special assessment
C77	B St. sewer	14th–20th	500	500	50	100	350	—	—	Private funds
C78	88th Ave. sewer	D–L	1,200	1,200	—	150	850	200	—	Spec. assess. 50%, rev. bonds 50%
C79	A St. sewer	90th–97th	1,000	1,000	—	—	200	600	200	Priv. funds 40%, spec. assess. 60%
C80	76th Ave. sewer	F–G	600	600	—	—	100	450	50	Special assessment
C81	43rd Ave. sewer	G–P	1,300	500	—	—	—	100	400	Spec. assess. 50% rev. bonds 50%
C82	57th Ave. sewer	C–X	4,500	500	—	—	—	—	500	Priv. funds 20%, rev. bonds 20%, spec. assess. 60%

Figure 6-2

costs which will be required to properly operate and maintain the new facility. Expected reductions in operations and maintenance costs will obviously be important factors in justifying high priority for proposed construction. It is essential that the capital improvements program be revised at least annually, especially so that it will provide up-to-date base information for the budget program and annual budget preparation of the water or wastewater utility.

Multi-Year Budget Program Preparation and Use

An effective budget which will provide adequate funds for capital improvements and the required operations and maintenance expenditures must be based on a long-range facilities plan, a detailed multi-year capital improvements program based on the long-range facilities plan, and accurate estimates of the resources required for operations and maintenance expenditures. The technical and fiscal planning for facilities construction is adequately provided by the long-range planning activity and the capital improvements program. Estimates of future operations and maintenance expenditures are included along with capital expenditures in the multi-year budget program.

The estimation of operations and maintenance expenditures four or five years into the future must take into account the capital improvements program construction over the program period, and include the anticipated operations and maintenance expenditures required for both existing and planned facilities. The budget program is thus simply an extension of the capital improvements program, in that the estimated annual operations and maintenance expenditures are added to the estimated annual capital expenditures to develop the total estimated annual expenditures for the water or wastewater utility. The annual capital expenditures will include annual debt service plus funds allocated to capital construction in each year, and the operations and maintenance expenditures for each year should be estimated on the basis of facilities which will be in operation in that year.

A complete understanding of the need for the use of a multi-year budget program must be developed within the utility governing body and among the utility administrative staff if a budget program is to be valuable to the manager and decision-making body of a water or wastewater utility. The annual budget should be a refined and detailed first-year segment of a multi-year program, and the budget program should provide a complete view of the future implications of budget year expenditures or of decisions to withhold certain expenditures in the budget year.

In the preparation of a multi-year budget program the first step should be a review of the many assumed constraints and conditions to be encountered during the program period: the programs, projects, changes in regulations and standards, and all other known factors which will require the expenditure of resources during the program period. These assumed conditions should be documented for consideration and guidance throughout the preparation of the budget program. The next step is the development of detailed estimates of the resources required to adequately operate and maintain the facilities of the water or wastewater utility and to accomplish the programs and activities of the utility. The resources normally required for meeting the objectives of the utility include, among many others, personnel, other utilities, chemicals, spare parts, tools and equipment, and consulting services of various kinds. An efficient system of records, reports, and resource estimation is essential for the reliable determination of the resources proposed for the budget program. This will include the number of man years for accomplishing specific tasks, the amount of electric power and natural gas required for specific activities, the amount of water used for annual operation and maintenance, the number of tons of various chemicals required for treatment, the historical use of and probable need for spare parts depending on age and condition of equipment, and the probable cost of routine and special consulting services and of all the other resources for which a need is anticipated in each year of the program period. In some cases it will be necessary to conduct special studies to determine probable changes in the resources required by the utility, or to develop reasonable estimates of the cost of operation and maintenance of a new facility expected to be in operation during the program period.

It cannot be emphasized too strongly that the determination of the unit amounts of the various resources must precede the estimation of dollar amounts for any of the program activities. The allocation of dollars to the program activities cannot be accomplished in a meaningful or reliable way until the real, nonfiscal resource needs have been established.

After the quantity resources requirements during the budget program period have been determined, it is a relatively routine matter to apply to those resources the appropriate projected unit costs. These unit costs should have been developed from national and regional economic trends and many other factors.

One of the important messages contained in this book is that budget programs and budgets must be based not on historical budgets, budget programs, or appropriations, but on detailed analyses and projections of needs and estimated costs and on records of performance and actual past expenditure data. The multi-year budget program, if conscientiously developed and utilized, will provide valuable guidance to the water or wastewater utility manager and to the governing body of the utility in scheduling improvements to the system; in planning and implementing changes in operations and maintenance; in planning for and scheduling increases or revisions in water or sewer rates, connection fees, and other methods of revenue production; and in providing improved management of not only the financial affairs but all activities of the utility.

Annual Budget Preparation and Control

The preparation of the annual budget for a water or wastewater utility should be relatively easy if adequate planning and programming have been a routine activity. The planning information necessary for budget preparation is available from the long-range facilities plan and the capital improvements program, and the required programming information can be found in the multi-year budget program previously discussed. The same basic principles apply to biennial budgets as well.

In general terms, the annual budget should always be the first-year segment of the multi-year budget program, but in an updated form based on current projections and records data. The budget program will normally have been completed several months in advance of the preparation of the annual budget, and only the application of current information relative to projected resource units and costs will be necessary to prepare the next year's annual budget. Obviously, certain situations will arise in which major programs or operational changes have required major changes in the budget program and the annual budget. When these changed conditions are applied to the annual budget, appropriate changes should also be made in the multi-year program to provide a review of the future impact of the new conditions.

The annual budget, like the budget program, must be based on the work program of the agency, the projected resources necessary to accomplish the work program, and projected unit costs based on economic and technological factors. It should normally be prepared in such a manner as to avoid the inclusion of excessive contingencies and also to avoid a conservatism that would lead to intentional underexpenditure throughout the budget year, i.e., overbudgeting. The financial reserves of the water or wastewater utility, whether operations and maintenance reserves, general reserve, bond payment reserves, or other reserve funds, should remain at reasonable levels. This will require a continuing effort to reduce excessively high balances to reasonable levels. During times of high reserve balances, it is possible and desirable to take these high levels into account in the budget preparation and to intentionally budget less than the anticipated expenditures. During times of low reserve balances, on the other hand, it is necessary to be conservative and to provide contingency funds to guard against the depletion of reserves below acceptable levels which could result in fiscal problems for the utility and the utility manager. Prudent fiscal management obviously dictates that the reserve fund balance levels must be examined periodically to determine whether these balances are adequate to provide the intended fiscal protection. Any reserve fund balance found to not provide the needed fiscal protection should be adjusted accordingly during budget preparation.

Successful budget administration requires that the manager of a water or wastewater utility exercise reasonable expenditure control during the fiscal year to make even the best budget and budget program prove to be effective. It is the responsibility of water and wastewater utility managers to keep informed of the status of expenditures on at least a monthly basis throughout the fiscal year, and to adjust expenditures, priorities, and programs as necessary to ensure maximum benefit from expenditures and to ensure that expenditures for high-priority activities take precedence over the expenditures for lower priority activities. Monthly expenditure status reports are essential for proper management control, and the use of such expenditure status reports for comparing projected annual total expenditures with the corresponding budget amounts can provide the budget and expenditure control necessary for effective fiscal management.

When the monthly expenditure status reports are provided to the utility manager they should be analyzed not only by the manager, but also by department heads or division heads. It is important in the review of these reports that the manager understand the departmental or divisional implications of expenditure limitations or revisions, and it is equally important for the department heads or division heads to have a similar understanding of the impacts on the total utility of departmental or divisional expenditure trends. The department heads are in a position to provide decision information to the utility manager for the purpose of determining which expenditures should proceed without restriction and which expenditures can or should be restrained for budgetary control purposes. These expenditure decisions obviously must be jointly determined if they are to best serve the water or wastewater utility.

In the projecting of annual expenditures from monthly expenditure status reports, it is essential that projection methods much better than straight-line methods be utilized. Actual resource needs for the remainder of the specific budget year should be estimated and then used to project the total anticipated annual expenditures by adding the projected additional expenditures during the balance of the budget year to the expenditures of the year to date. Straight-line projections obviously can be used for estimating expenditures for resources whose use does not vary significantly throughout the year, such as certain utilities, chemicals, and certain administrative functions. The development of reliable methods of projecting anticipated annual expenditures can assist greatly in providing successful budgetary control, and it is a responsibility of the water or wastewater utility manager to ensure that the best available projection procedure is utilized by his administrative and operational personnel.

The Value of Coordinated Planning, Programming, and Budgeting

One of the primary values of any coordinated planning, programming and budgeting activity is the realistic and improved information provided in the budget program and the annual budget. The long-range planning activity and the capital improvements program provide, in addition to financial information, detailed information concerning the scheduling of facilities construction and thus the approximate dates by which new facilities will be in operation and will require funding for operations and maintenance expenditures. The addition of operations and maintenance expenditure information to the capital improvements program fiscal schedule will provide the multi-year program for all costs to be faced by the water or wastewater utility during the four-, five-, or six-year budget program period.

With the correct use of a properly coordinated planning, programming, and budgeting system all major facilities construction can be scheduled realistically, any changes in water or sewer rates can be scheduled well in advance of the time needed, personnel changes and the training of personnel in new skills can be accomplished well in advance of need, and all functions throughout the utility organization can be better planned and scheduled than if such a system is not in use.

The business of the water or wastewater utility can thus be conducted in an orderly manner with a maximum of helpful information readily available to the manager, rather than in a way which requires the manager to indulge in a fiscal guessing game if no coordinated planning, programming, and budgeting system is being employed within the utility.

In addition, decision-makers of the utility can see the implications throughout the planning period or the program period of expenditures which may be recommended but not authorized. This is particularly important to the water or wastewater utility manager, since the rejection of recommendations for specific expenditures will often result in the need for additional funds at some time in the future. At such times as these the manager can immediately begin planning for the appropriate time to recommend those future year expenditures, and also begin the educational process aimed at justifying specific expenditures to the decision-makers to avoid the probable adverse consequences of forgoing or postponing these expenditures.

An additional advantage of a coordinated planning, programming, and budgeting system is the development of an effective system of analyzing the resources and costs involved in all the activities of the utility. Construction costs can be analyzed with respect to their impact on the operations and maintenance budgets in future years, and the costs of capital improvements can be compared to operations and maintenance expenditures which could be delayed, reduced, or eliminated with specific capital construction. With an effective planning, programming, and budgeting system the manager of a water or wastewater utility can routinely and continually keep in touch with the technological, physical, fiscal, and political impacts of all actions or lack of action by management or the utility governing body.

Emergency Planning

The planning of a water or wastewater utility cannot be complete unless and until adequate planning for the protection of people and facilities from the effects of various types of emergency or disaster has been incorporated into the planning activity of the utility.

The emergencies for which the water or wastewater utility manager should plan include both natural and man-caused disasters. Natural disasters include earthquakes, tornadoes and hurricanes, other severe wind storms, severe flooding, adverse winter snow and ice conditions, or extreme low-temperature conditions. Man-caused emergencies include strikes, riots, explosions, and fires as well as war-caused emergency conditions. In developing emergency plans it is important that the first concern be for the protection of the people in and around the utility facilities; protection of the facilities themselves comes next; after that, assistance should be considered for other persons or agencies in the area who may be subjected to and endangered by the same emergency situation.

The emergency plan, in addition to providing for the specific activities of personnel of the water or wastewater utility, should include provision for the necessary equipment and materials as well as the proper maintenance and operation of the equipment needed for emergency conditions. This would include various types of heavy equipment, fuel, water and food supplies, materials such as for sand bags and shoring needs, and a multitude of tools, including portable pumps, chain saws, jacks, and many hand tools.

A formal agreement or at least an understanding providing for mutual emergency assistance between agencies should be developed with the appropriate police, fire, and rescue agencies, as well as other utility and local and state government agencies. It is important that water and wastewater utilities cooperate to the maximum possible extent with public works agencies at the city, county, state, and federal government levels to provide maximum protection and benefit to the general public in the time of emergencies.

Summary

A coordinated system of long-range facilities planning; shorter-term programming of capital, operations, and maintenance expenditures; and annual budgets is essential to effective technical and fiscal planning and successful management of water and wastewater utilities.

The design, financing, and construction of facilities as well as the financing of all activities of water and wastewater utilities should be based on a continuing, effective long-range facilities planning activity.

A capital improvements program consisting of facilities construction projects tentatively scheduled for the first four, five, or six years of the long-range facilities plan should set forth the priorities for the projects, the scheduling of the construction projects, the estimated costs of each of the projects, and the sources of revenue to be used for financing each of the construction projects.

The addition of estimated operations and maintenance costs to the scheduled capital improvement expenditures or debt service costs from the capital improvement program will result in a multi-year budget program which will provide the utility manager with a total fiscal plan for several years into the future. Preparation of the annual budget is then easily accomplished by applying appropriate revisions to the first year of the multi-year budget program.

7

Rates and Finances

The fiscal condition and particularly the fiscal stability of a water or waste-water utility will determine to a large extent the success of the utility in providing adequate service to its customers. Although both water and waste-water utilities received significant financial support from the general tax base of communities many years ago, in recent years most water and wastewater utilities have been self-supporting, with needed revenues being almost totally derived from service charges to the customers of the utility.

The Utility Must be Fiscally Self-Supporting

Inherent in the capability of a water or wastewater utility to provide a high level of reliable service to its customers is the requirement that the utility maintain a sound fiscal base. Such a sound fiscal base will be dependent upon adequate revenues from the customers who use the utility services as well as from other sources, such as payments for fire protection, special assessments, water or sewer tap fees, and federal construction grants. For fiscal soundness it is also essential that capital, operations, and maintenance expenditures be wisely planned, scheduled, and controlled in accordance with sound budgeting and expenditure control procedures.

Most water and wastewater utilities historically received revenues from the general tax base of communities—wastewater utilities more than water utilities in recent years. In addition, it was common for many years that water revenues in excess of annual expenditure needs would be diverted to the financing of other city services such as street maintenance, drainage and flood control, parks, recreation, and other general city services.

The development of a self-supporting revenue system for water and wastewater utilities became essential to their success as utilities. The ability of these agencies to provide construction of new and replacement facilities when needed and to operate and maintain facilities and equipment adequately depended on sound fiscal policies dealing with revenues, financial planning, and expenditures.

Many wastewater utilities have drawn on ad valorem taxes and other city revenues for all or most of their annual revenues, but the Federal Water Pollution Control Act Amendments of 1972 mandated the development by these utilities of self-supporting revenue structures to ensure adequate financing of construction programs and the operation and maintenance of facilities into the indefinite future. The self-sufficiency requirements of the Federal Water Pollution Control Act Amendments of 1972 brought about changes in the philosophy of financing wastewater utilities, primarily in the form of the adoption of user charges which would provide the revenues required for operation and maintenance of the wastewater utility facilities, with reliance on ad valorem tax support only for construction purposes. The Clean Water Act of 1977 provided relief for certain wastewater agencies, permitting them to use ad valorem taxes to finance operations and maintenance activities if these taxes had historically been their method of financing and if certain other requirements could be met by the specific ad valorem tax system.

The importance of water and wastewater utility self-sufficiency becomes clear if one considers the fact that the safety and reliability of water supply and treatment and the safety and adequacy of wastewater treatment are vital to the public health and the environment. The protection of public health and the environment are much too important for officials to take any chance in having inadequate funds available to operate and maintain facilities for transportation and treatment of water and wastewater and the safe disposal of sludges.

Criteria for Setting and Evaluating Rates

The development and adoption of proper water and sewer service charges requires that these service charges meet certain criteria. These criteria include equitability among various classes of service and especially within each class of service; adequacy of the rate structure to enable the utility to accumulate sufficient funds to provide adequate service to the customer; and stability of the rate schedules, so that the rates will not be changed every two or three years or tend to rise and fall periodically. The water and sewer rates also should, to some reasonable degree, represent the actual cost of serving various classes of customers (residential, commerical, and industrial).

The determination of what is equitable in water or sewer rates is open to much argument, but in general a schedule of water rates or sewer rates should at a minimum be fair to all residential, commercial, and industrial customers of the utility. Residential customers who use large quantities of water and provide large quantities of wastewater into the wastewater utility sewer system obviously should pay more for their services than residential customers, who, say, use 50 percent as much water and contribute 50 percent as much wastewater to the sewer system. The cost of supplying and treating water and the cost of treating wastewater depend to a large degree on the amount of water and wastewater treated and otherwise handled. It would be inequitable for all residential customers, for example, to pay the same service charge even though one group of residential customers used twice as much water and thus received twice the water service and considerably more wastewater service from the respective utilities than another. Likewise, it would be inequitable for commercial or industrial customers with extremely strong wastes to be charged on the same basis for wastewater treatment as a residential customer whose waste is of considerably less strength.

The rate structures for water utilities and wastewater utilities obviously must be established so that adequate funds are brought into each utility to ensure the required level of service to its customers. This requires that the revenues derived from water or sewer rates along with other anticipated revenues must be at least sufficient to finance the operation and maintenance of existing facilities, to provide the funds to pay off any bonded indebtedness on existing facilities, to provide adequate funds for the construction of additional or replacement facilities when needed, and to provide the funds needed for the management and planning of the activities and facilities of the utility.

Stability in water and sewer service rates charged by water and wastewater utilities is important both from the standpoint of providing satisfactory service to the customers at a cost which the customer can properly anticipate, and for which the customer can budget, and also from the standpoint of developing a feeling of confidence in the utility on the part of the customer. If water or sewer rates are raised frequently or if they are raised and lowered frequently, the individual customer will develop doubts about the quality of management, fiscal responsibility, and fiscal security of the utility. This, in turn, can result in poor public relations and hence questionable public support for the activities of the utility. Some customers also may find it difficult to budget adequately to be able to pay their utility bills in a timely manner if the future water or sewer rates are in question. The development of an effective planning–programming–budgeting system will enable the water or wastewater utility manager to adequately plan any rate changes well in advance of need. Such rate changes can be held to a minimum frequency with suitable advance planning.

Two differing opinions exist concerning the need for water and sewer rates to represent the actual cost of service to individual customers. There are those who feel that the actual charge for service to a customer several miles from a treatment facility should take into account the capital cost of the additional length of distribution main or interceptor sewer required to serve him, thus penalizing the customer for being located some distance from the water or wastewater treatment plant. The opposing view of those who support a true utility concept is that service should be provided to each customer connected to the utility system at a price which is dependent only on the amount of water or wastewater service provided, regardless of location within the utility service area. This obviously requires that a customer with a strong wastewater will pay a higher charge than a customer with a weaker wastewater because of the higher cost of treatment of the stronger wastewater, and also requires that the customer who contributes much more wastewater for treatment or who consumes more water will pay more than the customer who contributes less and uses less. In the latter case each customer pays for the quality and quantity of service he receives, but not for his location of service.

For many years water and sewer rates have provided for quantity discounts so that customers would pay a lower unit charge for the larger quantities of water consumed or wastewater treated. This was in accordance with actual costs to water or wastewater utilities to provide such services, inasmuch as the total cost for treating water or wastewater does not increase in proportion to the quantity treated, but rather at a lower rate of increase. This results not only from the fact that annual debt service payments must be made by the utility regardless of treatment quantities, but also because certain costs such as personnel costs do not increase proportionately with the quantities of water or wastewater which require treatment. Within recent years the need to develop and implement extensive water conservation programs in water and wastewater utilities has resulted in the development and implementation of water rates and sewer service charges which not only would eliminate the large quantity discount, but would in fact penalize the large quantity users of water and wastewater facilities. The resulting inverted rate structure, which charges higher unit rates for larger quantities of water or wastewater, has in numerous communities replaced the more traditional rate structure which charges lower unit rates for larger quantities of water or wastewater. In addition to other water conservation techniques, the inverted rate structure has been found to be an effective method, perhaps the most effective method, of convincing customers that they should conserve water.

Within the service area of most water and wastewater utilities, there are some customers whose actual water consumption or wastewater contribution is so small that the furnishing of such service to these customers at normal rates would lose money for the utility. To avoid this situation most water and wastewater utilities provide for a minimum charge to be paid by customers regardless of the quantity of water they use or the quantity of wastewater they contribute to the sewer system. These minimum charges should be set at a level that is not unreasonable to the customer, but is high enough to cover the proportionate share of administrative costs, meter reading and repair, and other miscellaneous costs in addition to the cost of operations, maintenance, and debt service.

The Federal Water Pollution Control Act Amendments of 1972 included provisions which required that wastewater utilities charge their customers in proportion to the service they receive, with the determination of service received to be based on quantity and strength of wastewater. This requirement was found to be troublesome for managers of wastewater utilities which based the sewer service charge on the water service charge. In these cases, if the water rate structure included quantity discounts, as most water rate structures have, the large quantity discounts for wastewater service were determined to be contrary to the provisions of the Federal Water Pollution Control Act Amendments of 1972. It thus became necessary to base the sewer service charge on water quantity rather than on the water service charge. This obviously resulted in a more complex accounting and billing system because of the need to compute the water and the wastewater charges separately rather than merely use a percentage of the water bill for the wastewater charge.

The sewer service charge, to be completely equitable for customers, must be based not only on the quantity of wastewater treated, but also on the strength of the wastewater which receives treatment. As the required level of wastewater treatment increased to a high level of secondary treatment and even to advanced wastewater treatment, the proportion of total costs attributable to biological or other advanced treatment processes increased, thus placing a larger emphasis on the strength of the wastewater from a cost of service and charge standpoint. The obvious result was the development of wastewater service charges which placed increased emphasis on the strength of the wastewater, so that the customer whose wastewater required a larger expenditure of power, chemicals, and other services for the necessary level of treatment and sludge processing and disposal would pay a greater share of the total cost of the wastewater treatment process.

In some cases, certain wastes received into the public sewer system would require special treatment because of specific constituents in the wastewater or because of an exceptionally high concentration of certain constituents. Since these wastes containing stronger or special contaminants require the expenditure of larger amounts of money and other resources for treatment, it is only proper that the charges to the customers who provide such wastewaters should be higher than the charges to customers who contribute wastes of normal or relatively low strength.

Use of Ad Valorem and Other Taxes

A substantial number of wastewater systems have been totally or partially financed by ad valorem taxes almost since the first sewer system was constructed and placed in operation. The development of the utility concept for wastewater systems resulted in replacement of ad valorem taxes by sewer service charges as a source of revenue, but at the present time many small wastewater agencies and a few large agencies still rely on the ad valorem tax for financing not only construction, but also the operations and maintenance activities of the wastewater system.

Although ad valorem taxes have been used to some extent as a method of finance for water systems, their use has not been nearly as common as in the wastewater field. The realization that public water systems are utilities to be financed through direct service charges to customers became common long before wastewater systems were considered to be similar public utilities. There is still a widespread feeling that the treatment and transportation of wastewater are not in the same utility category as the supply, treatment, and distribution of potable water in a public water system.

An advantage to small water or wastewater agencies in using ad valorem taxes for financing the total programs of the agencies is the simplicity of the financing system, and the associated economies. By using ad valorem taxes a water or wastewater agency may be able to manage without the office space, equipment, and personnel required for billing, accounting and other administrative tasks in a user charge system. The ad valorem tax is normally collected by the county or city government for a relatively small charge each year, and the total cost to the water or wastewater utility of collecting needed revenues is a small fraction of the cost of a user charge system.

The Federal Water Pollution Control Act Amendments of 1972 discouraged the use of ad valorem taxes as a revenue source to pay operations and maintenance costs of wastewater utilities. Indeed, requiring that user charge systems be used by agencies which receive federal construction grants, knowing that probably no wastewater agency could refuse to seek the 75 percent federal grants, Congress in essence prohibited the use of ad valorem taxes to cover operations and maintenance costs. The Clean Water Act of 1977, as a result of considerable criticism from officials of wastewater utilities, contained provisions which provided for the continuation of ad valorem tax financing of operations and maintenance costs as well as capital costs for those wastewater agencies which had historically used such taxes for these purposes if certain conditions were met. These conditions included the requirement that the agencies show to the satisfaction of the USEPA Administrator that the ad valorem tax system charges customers approximately in proportion to the actual cost of providing service.

The Federal Water Pollution Control Act Amendments of 1972 also included industrial cost recovery provisions which required wastewater utilities who received federal construction grants to exact and collect special payments from industries connected to wastewater facilities partially financed by federal grant funds. According to the industrial cost recovery provisions each industry was required to repay their proportionate share of the federal grant over a thirty-year period, or sooner if desired. The Clean Water Act of

1977 placed an eighteen-month moratorium on the collection of these industrial cost recovery payments and also ordered the USEPA to conduct a study to determine whether the industrial cost recovery provisions of the 1972 Amendments were providing the benefits anticipated by Congress. The resulting study revealed what should have been expected, namely that the industrial cost recovery provisions would not result in the benefits or impacts expected by Congress. This is an example of how any type of utility charges which discriminate against one class of user rarely will result in more benefits than detriments for any user group. Regardless of the ultimate Congressional action concerning industrial cost recovery, this should be viewed as one representative sign of the type of unreasonable and unwise special constraints which water and wastewater utility managers can expect to face in the future.

Because of the strong feeling among members of Congress and other people throughout the United States who support and encourage federal regulation of local government, including the systems of charges of wastewater utilities, it is difficult to predict to what extent the Federal Government will force their regulation of water and wastewater utility revenue methods upon the managers and governing bodies of such agencies. Even though water and wastewater utilities should, for utility business reasons, utilize revenue bases which rely on user charges for revenue, it is important that utility governing bodies and managers be free to develop and utilize whatever revenue sources are best for the people they serve, rather than be forced to comply with revenue system constraints imposed by the Federal Government. Every knowledgeable person in the United States is completely aware that local government officials are considerably more capable of establishing and maintaining proper revenue systems than any elected or appointed federal officials possibly could be. Congress correctly feels that a water or wastewater utility should operate as much as possible as a utility, but the decisions concerning revenue sources and methods obviously should remain completely the prerogative of the local utility. Managers of water and wastewater utilities should routinely exert appropriate influence to prevent the Federal Government from interfering with this type of local decision.

Rate Equitability

The subject of equitability of water and sewer rates is important enough to warrant considerable discussion and thought. The generally accepted concept involved in setting the charges for water or wastewater service is that the customer should pay his proper share of the cost of providing the service, with no special consideration being given to special circumstances such as the distance the specific customer happens to be from the water treatment plant or wastewater treatment plant. For many years it has also been an acceptable practice to provide quantity discount rates for both water and wastewater service, since certain fixed costs of water and wastewater utility service make the unit cost of providing service less for treatment of larger quantities of water or wastewater. It has thus been common practice to provide a declining unit rate for water or sewer service as the quantities of water used or wastewater contributed increase.

The Federal Water Pollution Control Act Amendments of 1972 and the implementation regulations require that the use of such quantity discounts be discontinued for wastewater service, and it would appear that this requirement is contrary to the actual cost of providing service. The ultimate impact of this conflict is not at present known and may not be known for many years. In a later discussion on the use of inverted rate structures for water conservation purposes, additional consideration will be given to the wisdom of eliminating the quantity discount rates.

To a certain extent, water utility costs are related to peak flows, but much less so than in the case of wastewater treatment costs. While water supply systems normally include treated water storage, which usually alleviates serious peak demand impacts on the water treatment facilities, similar storage of raw wastewater during peak periods for treatment during off-peak periods is not commonly available in sewer systems. In some communities consideration has been given to charging lower sewer service rates for flows which are transported to the wastewater treatment facilities during the off-peak periods at night. Since wastewater treatment facility capacity must be designed for peak flows to a large degree, it appears reasonable that those customers who can hold their flows for release during off-peak times should receive a lower rate than those customers who release their flows for treatment during the peak periods. Even though the impact on water utilities is not quite the same as for wastewater facilities, the same logic may exist for some water utility customers who can take major flows during the off-peak periods.

In attempting to give adequate consideration to the subject of equitability of water and sewer service charges, it is necessary that the service charges system provide that such equitability exists within specific classes of user service. This requires that the rates apply fairly and equitably to all residential utility customers as a class of user, also within each of the commercial classes of customers, and within the industrial and institutional classes. Certain commercial customers will provide stronger wastewaters than others, and the additional costs which are incurred in treating those stronger wastewaters, if they actually occur and can be identified, should obviously be reflected in the service charge structure. This is necessary also for industrial users, taking into account both the strength of wastewaters received for treatment and any special contaminants in specific industrial wastes which would either interfere with the wastewater treatment process or would require additional or more expensive types of wastewater treatment. Since the level of treatment for water customers remains constant regardless of whether they are residential, commercial, or industrial, there is no similar need to distinguish among these classes of customer. For those occasional situations in which a separate or special water supply is provided to a customer, charges must reflect the actual cost of providing the special service.

In most communities the water and sewer service rates for customers outside the city have historically been set higher than for customers within the city. This higher rate, often amounting to a 50–100 percent increase over the in-city rate, has been intended to force the out-of-city customer to provide additional financial support to the water or wastewater utility in lieu of the ad valorem taxes or other revenue producing methods utilized within the city in which the out-of-city customer obviously does not participate. This type of charge structure has also been used to encourage annexation of outlying areas into cities. This differential in sewer service charges between in-city and the out-of-city customers was one of the service charge practices which was prohibited by the Federal Water Pollution Control Act Amendments of 1972 but could be permitted if the wastewater utility management could adequately document the additional costs incurred in serving the out-of-city customer.

To ensure that any significant additional costs of treating exceptionally strong wastes are paid by the contributors of these wastes, it is common for surcharges to be added to the normal wastewater charges for strong waste contributors. Similar surcharges to water or wastewater customers for large peak demands are not in general use, but should be considered. The use of surcharges for both of these situations is one additional method for ensuring equitability among water and wastewater utility customers. The surcharges will help to ensure that insofar as possible each customer will pay his fair share of the total cost of owning and operating the specific utility.

Tap Fees Versus Rate Increases

The continuing growth of an urban area imposes a fiscal burden on the citizens of that area in the form of financing the construction of expanded facilities needed by water and wastewater utilities in order to provide service to new residential, commercial, and industrial customers. Utility customers in many urban areas have been concerned about their moral and fiscal responsibility for financing expansions of their water and wastewater systems to provide capacity for new water and wastewater customers in the urban area.

Many arguments have centered around who is responsible for paying off the bonds and for financing the operations and maintenance costs of existing facilities, as well as financing the construction of additional facilities needed to meet the future demands on the utility. The responsibility for financing of existing facilities, both the annual debt service costs of past construction and the costs of operating and maintaining those facilities, would appear to be properly shared by the customers currently using the system and the future customers of the system.

For many years some water and wastewater utilities have included in their fiscal system a requirement for new customers of the water or wastewater utility to make an initial payment to the utility for the purpose of paying their share of the costs of past planning, administration, and construction to provide the facilities currently available to the new customers. This fiscal requirement enables a new customer to buy into the established water or wastewater utility, whose facilities were constructed with funds furnished by former and present customers.

The emphasis on growth paying its own way in a community and new customers buying into an existing utility system has led to the use of tap fees, plant investment fees, water or sewer connection fees, and similar charges, whatever they may be called. The terms may differ, but the buying-in concept is the same. These initial connection charges are intended both to provide revenues for paying off existing debt which was used to finance the construction of water or wastewater utility facilities, and to provide some of the funds for future expansion. The use of a water or sewer connection fee obviously also helps to keep the water and sewer charges stable in the face of new facilities construction necessitated by additional water and sewer customers. The connection fee, in addition to paying a share of facilities construction or debt service, should also cover the costs of pavement repair, administrative work, inspection of water or sewer connections, and any other costs incidental to new water or sewer connections.

Inverted Rate Structures for Water Conservation

As previously mentioned, rate structures for water and wastewater utilities have historically included quantity discounts intended to afford large quantity customers the advantage of the lower unit cost of operations and maintenance and lower capital allocation incurred by treating larger quantities of water or wastewater. This has been considered a fair and proper method of charging for water service because many costs such as debt service and personnel and equipment costs are fixed or not directly proportional to quantities of water treated and furnished to customers. The same reasoning is true of wastewater facilities and quantities of wastewater transported and treated.

In areas of the country which have experienced drought conditions or periods of water shortage, it has been found necessary for managers of water utilities to develop methods for conserving water and reducing the waste of water. In addition, with the development of energy shortages throughout the United States and around the world, it has also been increasingly important to reduce the amount of wasted water so as to reduce the total amount of energy required for treatment and transportation of water and wastewater, and also to reduce the amount of energy required for heating water in homes and in commercial and industrial establishments. Additional incentives for reducing water usage have been developed as a result of the increased levels of treatment required for wastewater and the need to decrease capital costs as well as operations and maintenance costs of wastewater facilities.

In an effort to provide effective water conservation, some water utility managers have developed an inverted rate structure which provides that the unit cost of the first given quantity of water supplied to a customer is lower than the unit costs of succeedingly larger quantities of water. Similar rate structures for sewer service charges could also be effective in encouraging water conservation, but federal regulations governing user charge systems may preclude suitable use of such an inverted rate structure for sewerage service. This inverted rate structure concept takes advantage of the financial impact on customers from the wasting of water. Experience has shown that with an inverted rate structure, people will continue to use water in generally the same ways and in substantially the same quantities as without the inverted rate structure, but with the inverted rate structure more emphasis in the home and in commercial and industrial establishments is placed on reducing the waste resulting from leaky valves, leaky toilets, indiscriminately filling bathtubs to high levels, and washing only partial rather than full loads of clothes or dishes.

The validity of the concept and the intended results of the inverted rate structure have been realized in practice. Specifically, it has been found that people are generally reluctant to pay significantly higher prices for increasingly larger quantities of water supplied or wastewater treated, especially if those larger quantities are the result of waste. It has been found that customers of water and wastewater utilities will continue to use as much water as necessary, but that on the other hand they will tend to reduce the waste of water in as many ways as possible.

One specific aspect of the inverted rate structure which must be analyzed carefully before implementation is the minimum charge for this type of rate structure as compared to the minimum charge for the normal rate structure. With an inverted rate structure system in use there will be a conscious effort by customers of the utility to reduce the total amount of water used. The result will be a total reduction in revenue which can be significant unless the minimum charge is revised to compensate for the water conservation program effectiveness. The minimum charge should be set at such a level as to take into account the anticipated reduced water usage and to compensate for the lower resulting revenues.

Examples of Water and Sewer Rates Actually in Use

As an aid to understanding both the similarities and differences in the actual rate structures of individual water and sewer utilities, some key provisions have been excerpted from the rate schedules of several actual utilities for presentation and comparison here. Connection charges have been omitted from the examples.

In analyzing the rate structure examples, special attention should be given to the minimum charges, the charge periods, the comparison of in-city rates to "outside" rates, the relationship of sewer rates to water rates, the inclusion of large quantity discounts in the rate structure, and the complexity of the rate structure.

The pertinent provisions for City A are as follows:

1. Residential and commercial water users are billed quarterly.
2. Users within the city pay a minimum quarterly charge (based on meter size) for usage up to 9,000 gallons per living unit; $0.60 per thousand gallons for the next 15,000 gallons; and $0.65 per thousand gallons for usage in excess of 24,000 gallons.
3. Users outside of the city pay a minimum quarterly charge (based on meter size) for usage up to 9,000 gallons per living unit; $1.10 per thousand gallons for the next 15,000 gallons; and $1.20 per thousand gallons for usage in excess of 24,000 gallons.
4. The in-city minimum quarterly charges, based on meter size, are shown in Table 7-1.
5. The out-of-city minimum quarterly charges based on water meter size are shown in Table 7-2.
6. Sewer users are billed quarterly at the time of water billings.
7. The minimum quarterly sewer charge, based on water usage up to 9,000 gallons is $5.00; all sewer usage in excess of 9,000 gallons of water per living unit is $0.40 per 1,000 gallons of water used during the first quarter of each year.

Table 7-1 / In-City Minimum Water Charges, City A.

Meter Size	Minimum Quarterly Charge
Less than 1 inch	$ 6.75
1 inch	8.85
1¼ and 1½ inch	13.80
2 inch	20.90
3 inch	41.00
4 inch	69.10
6 inch	149.50
8 inch	262.00

Table 7-2 / Out-of-City Minimum Water Charges, City A.

Meter Size	Minimum Quarterly Charge
Less than 1 inch	$ 13.50
1 inch	17.70
1¼ and 1½ inch	27.60
2 inch	41.80
3 inch	82.00
4 inch	138.20
6 inch	299.00
8 inch	524.00

8. The base sewer charges for commercial and industrial users are the same as for residential connectors.
9. In addition to the base sewer charges all commercial and industrial connectors are assessed a surcharge for combined biochemical oxygen demand and suspended solids in excess of 600 parts per million (5 pounds per 1,000 gallons of wastewater flow) at a rate of $0.08 per pound. Industries which contribute industrial wastes requiring special treatment will also pay an additional surcharge to be determined by the utility manager.

Of particular note concerning the water and sewer rates of City A are the following:

1. The out-of-city water customer is charged double the rate of the in-city customer.
2. Although the unit cost of water quantities from 9,000 gallons quarterly to 24,000 gallons quarterly is less than the unit cost for less than 9,000 gallons (minimum charge), the unit cost for quantities in excess of 24,000 gallons per quarter is greater than the 9,000–24,000 gallon unit cost, indicating a slight attempt to limit water use with an inverted rate structure.
3. The sewer service charge is identical for in-city and out-of-city customers, in compliance with the requirements of the Federal Water Pollution Control Act Amendments of 1972 and the Clean Water Act of 1977, which provide for comparable charges for comparable service regardless of location.

The comparable provisions for City B are as follows:

1. Residential and commercial water users are billed monthly.
2. Residential water users, whether within the city or outside the city pay a minimum monthly charge of $5.25 for up to 1,000 gallons and $0.75 per thousand gallons for water use in excess of the minimum 1,000 gallons per month.

3. Commercial water users pay a minimum monthly charge in accordance with Table 7-3, and pay a charge of $0.75 per thousand gallons for usage in excess of 1,000 gallons per month.
4. Sewer users are billed monthly at the time of water service billing.
5. Residential sewer charges are $6.00 per month per living unit.
6. Commercial sewer charges are 50 percent of the water charges or $6.00 per month, whichever is greater, if the user is receiving water service from the City. If the sewer user is not receiving water service from the City, the sewer charge is as negotiated with the City Council.
7. Industrial sewer service charges are based on flow, biochemical oxygen demand, suspended solids, and other strength indicators determined by the city to be appropriate, and the charges are as negotiated by the City Council.

Special note should be taken of the following:

1. City A water and sewer charges are billed quarterly whereas water and sewer charges for City B are billed monthly.
2. The minimum water charges for City A are considerably lower than those for City B.
3. Whereas City A has considerably higher water rates for out-of-city customers than for in-city customers, the rates for water service by City B are the same for all customers regardless of location.
4. Although City A has incorporated into its water rate structure a higher unit charge for large quantities of water use (in excess of 24,000 gallons per quarter), no such provision is included in the City B water rate structure.
5. Both City A and City B have sewer service charges which are the same for customers outside and inside the city.
6. The residential sewer service charges for City A are determined by water usage as indicated by water meter readings, whereas the City B residential sewer service charges are a set $6.00 per month. Commercial and industrial sewer service charges of City A have the same base as the residential charges, but a surcharge is applied for those flows which are of greater

Table 7-3 / Minimum Commercial Water Charges, City B.

Meter Size	Minimum Monthly Charge
Under 1 inch	$ 5.25
1 inch	10.00
1¼ or 1½ inch	19.00
2 inch	28.00
3 inch	56.00
4 inch	98.00
6 inch	219.00
8 inch	390.00

Table 7-4 / In-City Residential Water Service Charges, City C.

Meter Size	Monthly Service Charge
Less than 1 inch	$ 1.35
1 inch	2.25
1¼ inch	3.00
1½ inch	3.75
2 inch	5.25
3 inch	11.50
4 inch	19.50
6 inch	38.50

strength than residential flows. The industrial sewer service charges of City B, on the other hand, are subject to negotiation, as are the sewer charges of commerical connectors outside the city.

For a third city, identified as City C, the water and sewer rates are structured so that each customer pays both a monthly service charge based on the water meter size and a quantity charge based on the amount of water used, as follows:

1. The residential water rates for customers inside the city include monthly service charges as shown in Table 7-4. In addition to the monthly service charge, residential customers in City C must pay $0.83 per 1,000 gallons of water during the months from November through April. During the rest of the year the quantity charge is $0.83 per 1,000 gallons of water used for quantities not greater than the average monthly use during the November through April period, and $1.50 per 1,000 gallons for greater quantities.
2. Commercial rates inside the city are based on meter size and water usage; the usage rate for all quantities in excess of the minimum quantity for the specific meter size is $1.37 per 1,000 gallons. The minimum rates are shown in Table 7-5.

Table 7-5 / In-City Commercial Water Charges, City C.

Meter Size	Minimum Monthly Charge	Water Usage Included in Minimum Charge
Less than 1 inch	$ 5.50	1,000 gallons
1 inch	16.00	8,000 gallons
1¼ inch	21.00	12,000 gallons
1½ inch	33.00	20,000 gallons
2 inch	61.00	40,000 gallons
3 inch	117.00	80,000 gallons
4 inch	145.00	100,000 gallons
6 inch	(negotiated by city)	

3. The industrial water rates are all negotiated by the city.
4. For water service outside of City C, the rates for residential, commercial, and industrial customers are all 125 percent of the rates for customers inside the city.
5. The sewer rates for City C, similar to the water rates, are 125 percent of in-city rates for all customers outside the city. The monthly sewer charge, similar to the water charge, is composed of a monthly service charge based on the water meter size plus a monthly quantity charge based on the monthly water meter readings. The monthly quantity charge for the entire year is $0.78 per 1,000 gallons of water used during the period from November through April (winter usage). The monthly service charge, based on water meter size, is in accordance with Table 7-6.
6. For commercial or industrial customers, an additional surcharge is charged for wastewater which has a strength above that of a normal residential wastewater. The excessive strength surcharge is as follows:

$$S = Vs \times 8.34 \, [BC(BOD - 240) + SSC(SS - 255)] \times k$$

where S equals excessive strength surcharge in dollars; Vs equals monthly wastewater volume in million gallons; 8.34 is the weight in pounds of one gallon of water; BC is the treatment cost for biochemical oxygen demand expressed in dollars per pound; BOD equals biochemical oxygen demand strength in milligrams per liter; 240 is the allowed biochemical oxygen demand strength in milligrams per liter; SSC is the treatment cost for suspended solids in dollars per pound; SS equals suspended solids strength in milligrams per liter; 255 is the allowed suspended solids strength in milligrams per liter; and k is a constant to clear units.

In comparing the water and sewer rates of City C with those of Cities A and B, special note should be taken of the 25 percent higher water and sewer rates paid by customers outside City C; of the lack of a graduated rate schedule, balanced by the significantly higher charge for water quantities in excess of winter-period water usage; and the complexity of the schedule.

Table 7-6 / Sewer Service Charges, City C.

Water Meter Size	Monthly Service Charge
Less than 1 inch	$ 1.25
1 inch	1.50
1¼ inch	1.75
1½ inch	2.00
2 inch	3.00
3 inch	5.50
4 inch	9.75
6 inch	18.50

Impact of Water Conservation on Revenues

The need for water conservation is well documented in water and wastewater utility history and is generally understood by the public. Since the early 1970s the water and wastewater utility customer who has been alert to national and international affairs has been aware that there is a serious need to conserve not only water, but also energy, other physical resources, and money. On many occasions these water and wastewater utility customers have responded affirmatively to such needs, and have substantially reduced water waste.

Any significant reduction in the amount of water used in a public water system and the comparable reduction in the amount of wastewater contributed to a wastewater system will in most cases result in lower total revenues to the utilities. This revenue reduction normally will be proportionately greater than the corresponding reduction in operations and maintenance costs. In addition, debt service payments, scheduled over long periods of time, must be made annually regardless of any reductions in quantities of water usage.

The planning assumptions on which budget preparation and other activities are based, as well as the actual planning for a water or wastewater utility, can be expected to change when the utility has an effective water conservation program in effect. The assumed rate of increase in per capita consumption of water or contribution of wastewater must be studied and revised according to more recent experience with and during water conservation programs. Per capita water and wastewater quantities may change so much as to require utility planners to assume patterns of decrease in these physical quantities throughout future years, rather than the historical increase. Thorough study of the other factors involved in determining these quantities is obviously a necessary part of the planning activity. In some extreme cases, the reliable projection of future water supply and wastewater flows will be virtually impossible before two or three years of experience with the water conservation program have been absorbed by the planners. It is important that planning personnel become thoroughly familiar at the earliest possible time with all the data which can be made available concerning changes in the water usage patterns which result from water conservation programs. The counteracting fiscal impacts of effective water conservation programs must be addressed not only in fiscal planning, but also in all other types of planning. Not only do revenues tend to be reduced by water conservation; the scheduling of capital expenditures as well as annual operations and maintenance expenditures must also be adjusted according to the new projections of water and sewer service demands. Reduced water consumption will to a certain degree tend to reduce the annual expenditures as well as the income of a utility. Good utility planning will require reasonably good estimates of both income reduction and expenditure reduction.

A serious problem which can develop as a result of an effective water conservation program is an adverse public reaction to the rate increases which may result from conscientious efforts on the part of the consuming public to conserve water. It may be difficult or even impossible for a citizen to understand and accept the fact that his water bill or sewer bill has not decreased even though he knows that he has been conscientious in reducing his waste of water and total water consumption. It is important that an effective public information program be instituted to explain the reasons for changes in specific water and sewer rates and the higher water and sewer bills which may result from effective water conservation activities.

Water or sewer rate increases which have resulted totally or partially from water conservation programs can be expected to cause at least some negative water conservation reactions on the part of many customers. It is often difficult for even a conscientious citizen to feel that there is any justification for him paying more money for less service while he is protecting the water supply for other citizens who may continue their water wasting without incurring a significant increase in their water or sewer bills. Managers of water and wastewater utilities would be wise to study thoroughly all the fiscal, technological, psychological, and public relations impacts of water conservation programs. Energy conservation efforts as well as environmental activities will certainly result in similar impacts through the foreseeable future. Wise managers will be prepared for confrontation with these problems which they almost certainly must face.

Bond Issue Financing

For many years the great majority of the construction of water and wastewater facilities has been financed by bond issues, either by general obligation bonds or by revenue bonds. It has occasionally been possible for a water or wastewater utility to accumulate the required capital funds in advance of capital needs, but in recent history the bond issue has been by far the predominant method of financing facilities construction.

The concept of financing construction with bond issues which are paid off over twenty, thirty, or forty years is sound even though over those years of bond repayment or debt service payments many dollars are spent in interest payments. Since capital facilities are to be used not ony by present customers but also by future customers of the water or wastewater utility, many of whom are not yet born at the time of construction, it is proper that the payment for such facilities be spread over at least the early part of the lifetime of new customers. In this way the burden of constructing capital facilities is truly spread as much as possible among all those who use the facilities and receive the benefits from those facilities.

As a part of the fiscal planning of water and wastewater utilities, it is imperative that the utility managers develop and maintain a planning schedule of annual debt service payments, future capital expenditure needs, total bonded indebtedness and the indebtedness limit of the utility. With such a planning schedule, a utility manager can schedule future bond issues in such a way as to minimize the financial impact in any one year or any few years, even though the construction schedule itself may be sporadic. By scheduling new annual debt service payments to start at about the time older bond issues are paid off, a utility rate can be kept relatively uniform over a period of many years. The proper management of capital construction, bond issues, and bond payments is important enough for the water or wastewater utility manager either to retain expertise in such matters on his staff or to be able to utilize such expertise whenever needed on a consulting basis. The utility manager obviously must also have a reasonable working knowledge of these financing principles and how they actually work.

Utilization of Federal Grants

Regardless of an individual's philosophy concerning the utilization of construction grant funds from any level of government for water or wastewater facilities construction, it is a fact of life that for a considerable number of years the Federal Government and some state governments have provided such funds for water or wastewater utility managers to use in their construction programs. The use of federal grants for water utilities has been limited only to rare occasions, but wastewater utilities have had the federal construction grant program for a considerable number of years. Regardless of the degree to which federal construction grants have helped or hindered the water pollution abatement programs of wastewater agencies, such funds have been made available to local wastewater utilities, and they have been universally used in financing wastewater facilities construction. The probable extent of availability of this type of funding in the future must be included in the fiscal planning of both water and wastewater utilities if the planning is to be realistic.

The utilization of wastewater facilities construction grant monies under any of the various Federal Water Pollution Control Acts since the middle of the twentieth century has provided a significant amount of money, admittedly on a sporadic basis, for scheduling into the construction financing programs of wastewater utilities. The providing of large construction grants to wastewater utilities reached a climax in the Federal Water Pollution Control Act Amendments of 1972, with 75 percent of the costs of facilities construction being made available from the Federal Government. In many states additional funds, often amounting to 12½ percent of total construction costs, made it possible for local wastewater agencies to finance construction of facilities with only 12½ percent of the total construction funds being provided on the local level. Many arguments can be made for and against the federal construction grant program as well as certain state grant programs, but since the federal grant program is a matter of law it has been common to include anticipated construction grant monies in the preparation of schedules of financing for construction by wastewater utilities.

In addition to the construction grant funds which have been provided by the Federal Government to wastewater utilities, a considerable number of research grants have been made available to various wastewater utilities and some water utilities for the purpose of financing research, development, and demonstration projects. Although the funds made available for this important area of research and development have been woefully inadequate, the Federal Government, through the USEPA and certain other federal agencies, has provided some impetus to the nationwide program of developing better methods, better technology, and better facilities for the treatment of water and wastewater. Inasmuch as this type of funding has historically been in extremely short supply, it is imperative that managers of water and wastewater utilities continually search for funds for research and development and be ready to utilize such funds when they become available, almost at a moment's notice in many instances. Much of the research money available across the nation has been provided to universities and public or private scientific groups for the performance of research which in many cases could have been accomplished more effectively and more completely by water or wastewater utilities personnel. It is the responsibility of managers of water and wastewater utilities to influence and convince Congress and the officials of federal agencies that such research and development funds, or an appropriate portion of such funds, should be provided to water and wastewater utilities for this type of work.

There is probably no manager or other official in the water or wastewater utility field who is not painfully aware that the effects of federal grants include delays in planning, design, and construction; the requirement to conform with multitudes of restrictive conditions, many of which are unreasonable and some virtually impossible; and considerable increases in costs of planning, design, construction, and even operations and maintenance. In the scheduling of the financing of facilities construction, it is essential that these effects of federal construction grants be taken into account when the cost estimates for construction, operations, maintenance, and administration are being developed. When the cost of a specific construction project has been estimated, it is essential that the total estimated cost be increased by at least 20–30 percent to allow for the additional costs which will result from the many conditions and delays caused by the federal agency administering a grant for the construction project.

For as long as federal construction grants or other types of financial grants are available to water or wastewater utilities, it will be necessary for the managers of these utilities to include consideration of such anticipated grant funds in their fiscal planning and scheduling as well as in the review and establishment of rates and connection charges to be assessed against their customers.

Sound water and wastewater utility management dictates sound fiscal planning and management, so all the rates, charges, and financial considerations mentioned here must remain high on the priority list of management tasks performed on a routine basis. In the study and development of water and sewer rates, the manager of a water or wastewater utility should rely on a consultant who is experienced in rate studies and rate setting. In addition, throughout his routine daily activities the utility manager must continually concentrate adequate staff, consultant, and personal expertise on all aspects of the financial affairs of his utility.

Summary

Inasmuch as the fiscal condition of a water or wastewater utility will govern the quality of service to the customers of the utility, it is essential that the utility manager develop and maintain an effective fiscal planning program to assure suitable rates and charges as well as adequate finances for the utility. This is particularly important to water and wastewater utilities because by their nature they are, or at least should be, financially self-supporting.

Water and sewer service charges should be established in such a way as to ensure fairness and equity as much as possible within the various classes of users, and also between classes of users.

Water and sewer service rates should be maintained at such levels as to ensure that adequate funds will be received by the utilities to provide the level of service to their customers which they desire and which is required to protect public health and the environment, as well as to comply with applicable laws and regulations.

Managers of water and wastewater utilities have an obligation to resist efforts of the Federal Government and state governments to dictate the manner in which the utilities obtain their revenues. Under no circumstances can Congress or a state legislature be more capable than the utility governing body of determining the types of revenues which will best serve the utility or its customers.

The thoughtful consideration of equitability of water and sewer service charges, the use of water and sewer connection (or tap) fees, the many fiscal consequences of water conservation programs, the proper use of bond issues and other financing methods for facilities construction, and the utilization of federal or state construction grants-in-aid must govern water and wastewater utility managers in their fiscal planning for their utilities.

The study of equitability and adequacy of water and sewer service rates and charges should be entrusted only to consulting firms and individuals who have a reputation of proven expertise in this highly specialized activity.

Automation, Data Processing, and Computer Process Control

The manager of a water or wastewater utility has available many opportunities to reap significant benefits from the use of automation in the activities of his utility. In recent years it has been found that automation can bring significant benefits to almost any type of governmental, industrial, or commercial activity where it can perform a function better, faster, or at lower cost. The wise utility manager will always keep in mind, however, that for every group of success stories concerning the use of automation there is at least one dismal story of failure. It is thus imperative that water and wastewater utility managers become familiar with both successful and unsuccessful applications of automation, and also be completely aware of the reasons for failure in the unsuccessful cases.

Automation Possibilities

Two areas in which the use of automation is easy and effective are record-keeping and reporting. The advent of computers and various types of automated equipment has made it possible to maintain thorough and concise records of virtually every conceivable type in electronic form, and also has made it possible for reports of many types to be delivered where and when needed in a fraction of the time previously required.

A common tendency among managers new to data processing, including water and wastewater utility managers, has been to develop systems of records and types of reports for the purpose of utilizing the capabilities of the automated equipment, rather than to utilize the available automated system to meet specific records and reporting needs. The wise utility manager will continually monitor the types of records and reports which are being produced, to ensure that only those records which are useful management or information tools are maintained, that only those reports which can be beneficially used by management, the policy-makers, or the general public are developed, and that the distribution of copies of reports be limited to those individuals who have a need to know the content of the reports. This is an important word of caution to water and wastewater utility managers: use computers and data processing equipment only where benefits are to be derived, not merely wherever somebody in the organization has found another way to utilize the equipment or system.

One of the important benefits to water and wastewater utility managers which has been derived from the use of computers has been the improved information retrieval systems which are now possible. As technology has developed in the water and wastewater fields, the volume of information needed for adequate planning, design, construction, operation, maintenance, financing, and administration of the many activities of the utilities has increased significantly. Without the use of certain types of computer systems it would be virtually impossible to store such volumes of information and then to successful retrieve the information when needed. A water or wastewater utility may have in its warehouse several thousand spare parts which must not only be cataloged and easily located, but must also be entered into a record system which will ensure that orders for additional spare parts can be processed in a timely manner whenever a prescribed reorder point has been reached. For a warehousing and purchasing system to operate satisfactorily, it is necessary that spare parts and materials which for any reason have become obsolete can be purged from the inventory system and disposed of in an officially prescribed manner. Additional information concerning the maintenance of various types of equipment and facilities, availability and qualifications of specific personnel, and many other types of management information can be stored and retrieved rapidly and appropriately with a computer system, something which could be accomplished only with extreme difficulty and with the expenditure of much personnel time in the case of large utility organizations before the advent of the computer.

One of the applications of computer and data processing equipment which results in a significant reduction in the amount of work required and accomplishes the task in a small fraction of the time otherwise required is the performance of the large quantity of engineering, laboratory, and scientific computations which must be accomplished within water and wastewater utility organizations for both routine activities and special projects and programs.

The faster computations make it possible for management and operational decisions to be made with current information rather than information which may be days or even weeks old. It is also possible to perform on a routine basis computations which previously would only be performed on a demand basis when the expenditure of personnel time and funds for manual computation was organizationally possible and could be justified because of the high-priority rating of a given project or program.

The monitoring of systems and processes has been possible with automation for quite a few years, and has enabled managers of water and wastewater utilities to obtain, process, and effectively use information in ways which were not possible decades ago. In water and wastewater treatment processes, the monitoring of pressures, temperatures, flow rates, densities, chemical and physical properties of fluids, and numerous other characteristics of the treatment process is routinely accomplished with ease. Overheated equipment, equipment failures, high water levels in chambers, and surcharging gravity sewers can be identified and reported to operations personnel instantaneously so that corrective action can be initiated virtually as soon as a critical situation has developed. Similar types of monitoring can provide data for system control or allocation of charges much more rapidly and less expensively with automation than with manual sampling and analysis, which was the method necessarily utilized for many years.

The utilization of computers for the control of treatment processes and of various facilities within water and wastewater utility systems has been made possible by the many years of successful development and application of computer process control in industry. The computer control of industrial processes was developed for the understandable purposes of reducing costs, increasing production, developing improved processes, and product quality control, as well as for avoiding critical process disruptions in the event of strikes or sabotage. The employment of computers for controlling water treatment or wastewater treatment processes or systems has resulted in considerable success in both types of utility. With the technological improvements of recent years in the manufacturing and installing of sensing devices, telemetering, and computer equipment systems, it has been possible for processes in water and wastewater treatment plants to be controlled virtually from the point where the water or wastewater enters the treatment facility to the point where the finished product is discharged.

The numbers and types of employees required to provide adequate control of the treatment processes has changed considerably as the result of the use of computer process control. The major changes have included a reduction in the total number of employees, a marked reduction in operations personnel, an increase in maintenance personnel, and, of course, the need for computer process control operators. The need for additional personnel as well as other resources to provide the necessary operation and maintenance of the process control equipment has lessened the personnel reduction benefits resulting from the use of computer process control, but in addition to the net reductions in personnel costs, which is significant, the many other advantages of computer process control would be worth the investment of resources even if no personnel reductions were possible. The computer process control of a water treatment plant or a wastewater treatment plant makes possible the instantaneous adjustment of valves, gates, chemical feed equipment, and other equipment to reduce expenditures of electric power, chemicals and other resources in response to changes in conditions of flow, concentrations of organic or chemical impurities, demands for oxygen, biological organisms or chemicals, or other process changes which routinely develop within the treatment plant.

Computer Process Control in Water and Wastewater Utilities

For many years in private industry, a considerable number of industrial processes have been controlled by computers. Some of the industrial processes which have been computer controlled have involved hydraulic and pressure control, temperature sensing and control, fluid density and viscosity control, and general piping and pipe flow considerations. It was a natural transitional development for some of the process control systems used with industrial hydraulic processes to be applied to a significant number of similar processes which are found in water and wastewater system facilities.

Within public water distribution systems there have been many successful applications of computer equipment in monitoring pressures and flows, as well as successful applications in also controlling flow distribution within water distribution systems with computer equipment. Similar applications have also been successful in both gravity flow and pressure sewer systems and appurtenant lift station operation. The application of computer control to wastewater treatment processes, even though relatively new and not widely attempted, has also been found to be successful when properly designed, operated, and maintained. Since computer equipment has sucessfully provided improved monitoring in wastewater treatment plants, and since computer process control has proved effective and achieveable, there is every reason to believe that computer control of wastewater treatment processes will be much more widely used in future years (Figure 8-1).

The feasible applications of computer control in public water systems include balancing of flows and pressures within the water distribution system, control of flows into and out of water distribution system storage tanks, control of pumping facilities, and partial or total control of many or all other processes within a water treatment plant. Among the processes which should

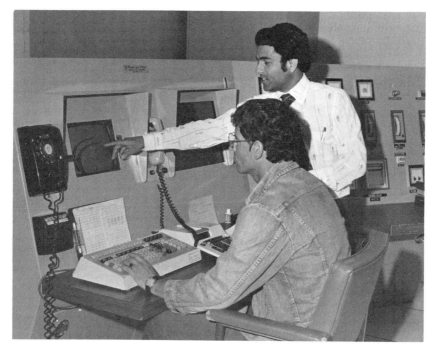

Figure 8-1 / "Nerve center" of the process control computer system of a major wastewater treatment plant.

be given serious consideration for computer control are chemical feed dosages and rates, chemical mix time, flocculation time, filter backwash frequency and duration, flow distribution to specific treatment units, and various flow rates. In the interest of energy conservation as well as the obvious accompaning economy, it is important to a water utility manager that insofar as possible pumping into storage, especially elevated storage, be done during off-peak electric power usage times. The application of computer control to this type of pumping to storage and similar operations should result not only in considerable energy savings, but also in reduced pumping costs and overall economy.

Within a wastewater utility system, the consideration of desired uses of computer control should include the control of pumping periods and scheduling at wastewater lift stations, the control of flow splits to various treatment units within treatment plants, the control of return activated sludge and waste activated sludge in secondary treatment facilities, chemical feed control in sludge processing facilities, and the control of chlorination rates to meet chlorine residual and bacterial concentration limitations as well as to minimize costs. Computer control can also permit maximum use of the capacities of large interceptor sewers by diverting flows from sewers running near capacity to those with more available capacity. For managers of water and wastewater utilities these few applications should serve only as representative examples of the possibilities of computer control.

Data Processing

The use of data processing equipment can provide the manager of a water or wastewater utility with improved records and reports of virtually all kinds. These include personnel records and reports, payroll records and check processing, operations and maintenance records and reports, purchasing and inventory records, and many other types of records and reports.

In most water and wastewater utilities it has been found that the use of data processing equipment for a variety of applications in the accounting system, for purchasing, and for the processing and payment of bills can result not only in improved accounting efficiencies, but also in staff reductions and accompanying reduction of costs to levels otherwise unattainable.

Most of the larger water and wastewater utilities routinely make use of data processing equipment for administrative activities, including those already mentioned—usually accounting, personnel records and reports, payroll, purchasing, and inventory control. Managers of water or wastewater utilities which do not utilize data processing equipment for these and similar activities should investigate the need for and benefits from data processing in such activities. If the need and prospective benefits appear to be present, further investigation should determine the costs involved, what steps should be taken to convert to data processing, and which of the activities of the utility should be so converted. Managers of water or wastewater utilities which do utilize data processing for these or other activities likewise should continually review the data processing applications to determine whether they are cost-effective and investigate whether other data processing application possibilities should be considered.

Most complex or lengthy engineering and scientific computations have been found to be much more rapidly and reliably performed with the use of computer equipment than by any other computation method. Certain engineering computations have been found to be particularly well suited for computer equipment. These include hydraulics computations and analyses, structural analyses, economic computations, earthwork quantities computations, and complex surveying problems among many others. Many types of laboratory computations and other scientific types of computations have also been found to be accomplished satisfactorily and rapidly with computer equipment. The transforming of contaminant concentration figures into total weight or total quantities is easily performed with computer equipment.

An example of this type of computation is the conversion of concentrations of suspended solids or biochemical oxygen demand in milligrams per liter into total pounds or tons of suspended solids or biochemical oxygen demand for a day, week, month, or year.

The use of data processing equipment in the implementing or improving of a preventive maintenance program will ensure faster, more reliable, and less costly preventive maintenance than could be provided otherwise. The difference between a manual system and a control system which utilizes data processing equipment should be obvious. The scheduling of literally thousands of inspections and maintenance tasks on hundreds of pieces of equipment, and the recording of all information concerning the pertinent corrective maintenance and replacement of parts are complex and time-consuming if done manually. Coordination of the spare parts inventory and purchasing control system with preventive and corrective maintenance activities cannot be adequately accomplished without the use of data processing equipment. The value of an automated preventive maintenance control system and related activities can probably best be appreciated by realizing how important it is to program specific personnel, tools, and spare parts to be at a specific location at a specific time to avoid equipment failure which could result in costly repairs or replacement of equipment and an interruption of the treatment process. Effective coordination of a preventive maintenance program with inventory and purchasing control of spare parts and other related activities can provide many benefits which the utility manager can recognize and appreciate. The additional coordination of a treatment process and equipment operation with the preventive maintenance program can reduce to a minimum the interference with operations caused by maintenance of equipment. Again, it should be obvious that with a treatment facility of significant complexity this coordination can be provided better with computer control than can be provided manually.

The use of data processing equipment in routine cost analysis and expenditure control enables the utility manager to receive expenditure and cost reports within only a few days of the end of an accounting period. This provides the manager and his staff personnel an early opportunity to analyze the expenditures for the specific time period and to project probable costs for the blance of the budget year with the expenditure of minimum time and minimum effort. This enables the manager to have a clear picture of the present time budget and expenditure status as well as a reliable projection of the year-end status. Not only is this capability of value to the water or wastewater utility manager; it is actually essential for him to have available for review by his key staff personnel and himself the most up-to-date expenditure information as well as the capability to analyze this information rapidly and often. Changing operations and maintenance conditions, cost fluctuations, and the many other expenditure variables encountered throughout a budget year make data processing of cost and expenditure information essential to effective budget and expenditure control.

Each of these data processing applications as well as the many other possible applications of data processing should be considered carefully for utilization by water and wastewater utility managers so that the maximum amount of information can be recorded and retrieved as rapidly and efficiently as possible, and so that the various engineering and scientific computations, preventive maintenance system, inventory control, purchasing, warehousing, and other activities can be improved to their maximum possible extent. Figure 8-2 shows a number of data processing functions being performed concurrently at a major wastewater utility data processing activity.

Time Sharing Versus Lease or Purchase of Equipment

Some of the best advice which can be given to the manager of a water or wastewater utility in connection with the development of a computer system is to exercise extreme caution when considering the use, purchase, or leasing

Figure 8-2 / Multiple data processing functions at a major wastewater utility.

of computer equipment. For many water or wastewater utility activities it is necessary to purchase and install equipment before the utility can function properly. This is not the case with computers. It is wise for the appropriate manual systems to be developed and tested before serious consideration is given to the best method of employing computers for these activities. This is true for all systems for which consideration may be given for automation. Only after the manual systems have been tested and are working satisfactorily should decisions be made concerning which systems are to be converted to computer use and whether computer equipment is to be purchased, leased, or used on a time-sharing basis. Each system or activity should be studied individually as well as in concert with other related activities, and the decision for each activity should be based on the results of those studies. The decision to use computer equipment for a specific activity should be made only when the results of the studies definitely indicate savings in time or cost, substantially improved information retrieval, improved computations or other similar substantial benefits.

Several time-sharing computer services are usually conveniently available for use by water and wastewater utilities, and it is normally wise for the utility manager to pay for time sharing until computer usage by the utility has developed to the point where leasing or purchase of equipment is feasible and economical. As a general rule, the acquisition of in-house computer equipment before the benefits of such in-house equipment have been well proven will be premature and probably regretted.

Staffing for Computer Activities

The many decisions concerning the development of systems and the use of computers for process control or data processing should not be made without considerable contribution to the decision-making process by an adequate staff of computer or data processing personnel. It will not be necessary to have a large number of such personnel for the required decisions, but it is imperative that the responsibility for specific decisions be borne by personnel who actually possess the needed computer expertise.

For treatment facilities in which the process control will be provided by computers, it is essential that process control operators be hired and appropriately trained early enough to ensure that they are fully qualified to initiate the necessary process changes with the appropriate computer commands from the beginning. Although a thorough knowledge of the treatment process is not a requisite of the process control operator, a reasonable knowledge of the treatment process obviously will be beneficial to both the operator and to the treatment process.

To maintain the computer process control system in a condition of adequacy, it is necessary for the utility manager to have available the services of a process control engineer who will be responsible for maintaining and revising the software and hardware portions of the system so that it will always be capable of responding promptly and correctly to changes in treatment requirements, flow, pollutant characteristics, and other factors. Whether such expertise is provided by a staff person or by a consultant will depend on many factors, including the needs, constraints, and wishes of the utility manager.

If the utility manager is to take advantage of computer equipment for engineering or scientific computations, it will be necessary for the appropriate computer programs and qualified personnel to be available. For this use, as for other uses, there is the question of purchasing computer programs and hiring consultants or staff employees to maintain them as opposed to retraining existing staff in program development and maintenance.

It is necessary for the accounting personnel, management, and other employees who will be involved with a computer system to understand the limitations and possibilities of computer data processing if the data processing needs of a water or wastewater utility are to be met satisfactorily. A computer programmer who is experienced in data processing applications must be available to the agency either on staff or as a consultant—preferably on staff. The decision as to whether accounting personnel or specialist data processing personnel should work directly with the computer will depend on the size and type of organization of the accounting and data processing activities within each utility, as well as the use intended for the various personnel, the accounting activities and functions, and the many administrative records and reports of the utility.

What to Expect from Computers

It is important that water and wastewater utility managers fully understand that computers cannot perform miracles, but like all other equipment, have definite limitations. It also is important for utility managers to understand that faulty or inadequate data fed into a computer will result in faulty or inadequate data coming out of the computer. The phrase "garbage in—garbage out" is descriptive of many computer system problems and should never be disregarded by users of computer equipment. Although most knowledgeable persons are aware of these computer facts of life, it is wise for utility managers to constantly be on the alert to avoid disregarding them.

The manager of a water or wastewater utility can expect from a computer system improved information gathering, storage, and retrieval, but only if the computer system has been properly designed, maintained, and updated, and if it is based on an effective manual system. Improved records and information systems can and should be expected from a computerized system, and reports from a computerized reporting system also should be much better than manual reports in terms of volume of information, thoroughness and accuracy of information, and rapidity in printing and releasing. If these results are not realized from computer applications, the manager is obligated to determine why.

The experience of many years has shown that computer computations are faster and more reliable than manual computations, especially in the case of complex computations. Simple computations, in most instances, are more properly and efficiently handled with noncomputer equipment, but complex, time-consuming computations should usually be performed by computer equipment. The utility manager should be aware of the types of computations being handled by computers as well as those computations handled by other equipment, and should ensure that his personnel use good judgment in choosing the proper type of equipment for specific types of computations.

Experience in the field of computer process control, especially in the control of industrial processes, has proved that the utility manager can realize considerably improved process control in water or wastewater treatment plants and other facilities by making use of computers. It is essential, however, that the process control system be properly designed, maintained, and revised as conditions warrant if it is to provide the desired benefits. The sensor equipment, the telemetry system, and the computer system all must be properly fitted into the control system, and all portions of the system must be adequately maintained if satisfactory process control is to be accomplished.

In addition to utilizing total or partial process control by computers, it is often desirable to provide for process monitoring in those cases in which actual process control is either not desired or not possible with computer equipment. For the purpose of monitoring of a water or wastewater treatment process as with process control, the sensor equipment, the telemetry system, and the recorders all must be properly fitted into the system and properly maintained. Such computer monitoring systems should be so designed that whenever conversion from manual control to computer process control is desired it can be accomplished with a minimum of conversion effort and cost.

The intent of this chapter is to make the water or wastewater utility manager aware of the many opportunities which exist with the use of computer systems, and to also caution the manager that his consideration of computer equipment for any phase of utility operations, maintenance, or administration, as well as any resulting decisions, must be based on the best available input from computer personnel as well as the operations, maintenance, and administrative personnel of the utility. To base decisions on less than the best possible joint expertise is to court failure.

Summary

The use of electronic computers in most activities of a water or wastewater utility will bring substantial benefits to the utility if well planned and properly managed.

Computers should be considered for use in the many data processing activities of water and wastewater utilities as well as for monitoring of treatment facilities and the water distribution systems and wastewater collection and interceptor systems. Actual computer control of water or wastewater treatment processes has been found to result in significant savings in the costs of personnel, energy, chemicals, and other resources.

Throughout the early years of development of a computer system for a water or wastewater utility, it is prudent for the utility manager to organize the various systems for time-sharing. Only when computer use and experience have progressed to the point where in-house equipment warrants consideration should the utility manager seriously embark on computer acquisition. At that time it is essential that the decision to lease or purchase be accompanied by planning for the retention of personnel of suitable expertise to ensure maximum benefit from in-house computer equipment.

It is important that water and wastewater utility managers become thoroughly familiar with the many ways in which computers can improve the administration, operations, and maintenance of their facilities and activities. It is equally important for the utility manager to understand the limitations of computers and the need to precede each computer application with a detailed study and evaluation of costs and benefits.

Legal Concerns
of the Utility Manager

The manager of a water or wastewater utility can be confronted by many types of legal problems and legal liability at any time during the conduct of normal utility activities, and it is imperative that at all times he have available expert legal advice and legal services on a continuing and reliable basis. Some of the larger utilities have a legal staff consisting of one or more attorneys who have extensive experience in the water or wastewater field and whose sole practice is with and for the utility. Many other utility managers may not have available the required expert legal advice in the form of a staff member, and these managers must ensure that they obtain the required legal services for their utility on a continuing and reliable basis in some other manner. They probably will negotiate with one or more attorneys who have appropriate water or wastewater utility experience, to serve as legal consultants to the utility. Whether in the form of staff attorneys or legal consultants, the legal services provided to the utility manager must be available when needed.

The Range of Legal Considerations

In the day-to-day conduct of the business of his water or wastewater utility, the manager will often be confronted with a variety of legal questions and legal actions which arise in connection with construction and service contracts; purchasing agreements; damages resulting from the activities of utility personnel; condemnation actions related to the acquisition of land, rights-of-way, or easements; water rights; and other types of legal questions and concerns too numerous to mention.

It is usually wise for utility managers whose activities must be conducted in accordance with western water law to retain the services of a specialized water rights attorney with extensive experience in water law and water rights. Questions which arise concerning water rights should be referred to the water rights attorney. A review by this attorney of utility activities for the purpose of advising the manager as to the probable legal implications of various alternative actions should be a routine procedure for those water or wastewater utilities which are constrained by water rights and water law.

Another utility activity which requires close coordination between the utility manager and his legal counsel is contract administration. Whether the contract is for services, purchases, or construction, it is essential that legal advice be sought and used in the preparation or review of the contract, the bidding procedure, and awarding of the contract, as well as whenever it appears that any party to a contract may be close to a breach of the contract.

The actions or negligence of employees or the failure of some part of a water or wastewater utility facility can result in legal liability because of damage to property, injury to persons, or nuisances.

The utility can often negotiate the acquisition of property, easements, or rights-of-way for facilities, but many times it must resort to condemnation action in the courts. These actions all obviously require guidance from legal counsel.

In the past decade or so many federal and state laws have placed increased legal burdens on the water and wastewater utility manager. To comply with the many laws and resulting implementation regulations, some found to be in conflict with others and certainly most confusing, it is essential for the utility manager to rely on his legal counsel for guidance, interpretation, and representation.

These few general examples of the types of legal problems which can face the water or wastewater utility manager should, at a minimum, make him appreciate the necessity for reliable legal counsel services in his daily routine as well as in special projects.

Legal Concerns of the Water Utility Manager

Prior to 1976, when Congress enacted the Safe Drinking Water Act, the managers of water utilities were not faced with unreasonably stringent federal laws or regulations controlling levels of treatment and required quality of drinking water. The water utility manager also was not faced with severe penalties in the event that the product of his water utility system did not measure up to predetermined standards.

During the many years of public water utility operations prior to 1976 the U.S. Public Health Service standards for public water supplies and revised versions thereof served as the water supply quality standards for public water systems across the country. Many states had also adopted regulations or standards which provided guidance and a reasonably sound level of legal constraint for water utility managers for many years. Public water supplies with only few exceptions were safe and adequate for most of those many years. Public health problems in a few locations in the early 1970s brought about a vociferous insistence from some segments of society for federal intervention into the construction, operation, and maintenance of public water supply systems. Water utility managers have since been faced with stringent requirements, which in many cases are admittedly desirable to a large extent, but like most federal constraints are not appropriate in all locations or for all water supply systems.

The primary function of a water utility is to operate and maintain safe, adequate, reliable, and economical water supply, treatment, and distribution facilities. In fulfilling this obligation to the public, the manager, the governing body, and the employees of a water utility can face serious legal problems. If any of the water utility employees perform their duties in a haphazard manner which results in injury or impairment of health to a citizen, the utility as well as its governing body, management personnel, and other appropriate personnel could be confronted by legal liability and action in a court of law. If a situation should develop in which a health hazard can be traced to improper filter operation or chemical feed control, or if it is found that a new section of water main has not been properly disinfected and has caused health problems or serious illness the utility can be faced with legal action. This legal action can be brought against not only the utility, but also the governing body, management personnel, or individual employees of the utility.

The planning, design, and construction of water utility facilities will be accomplished primarily in accordance with technological, fiscal, regulatory, and political factors, but the legal aspects of the location, quality of construction, and operation and maintenance of facilities must be considered as well. Planning must take into account the possibility of law suits arising from alleged nuisances, condemnation action for acquisition of easements or land purchase, and other legal problems which could result from the construction of facilities or the significant alteration of water utility activities. The design and construction of facilities, likewise, must be accomplished in such a way as to avoid as much as possible any infringement on the rights of citizens who reside or travel near construction sites or who live in close proximity to treatment facilities. The design and construction of facilities must be accomplished in a way which provides suitable consideration of features which will minimize conflicts and possible legal action during the period of construction as well as during the future years of operation of the facilities. With respect to the administration of construction contracts, it is essential that the construction documents be carefully prepared so as to minimize the necessity for legal interpretation of the plans or specifications as well as the other contract documents. Extra effort and funds expended at the time of design and contract award will be a small price to pay compared to the time and funds which could be expended in each case of serious construction contract misinterpretation or disagreement. The costs of such contract disagreement can be extremely detrimental to the water utility regardless of whether the disagreement is ultimately settled by negotiation or by court action.

A continuing legal concern for water utility managers is their personal liability and the liability of the utility as the result of acts of negligence of employees during the performance of their duties. In each case of injury to a person or damage to property resulting from an overt act of a water utility employee, the manager of a water utility faces the possibility of a law suit or a negotiated settlement, often involving large sums of money. It is essential that the water utility manager ensure that all employees of the utility are fully aware of the safe and proper ways of performing their duties, the types of activities and actions which are forbidden, and the probable legal results of improper actions on the part of the employees.

In addition to possible liability for the negligence of employees, a water utility can also be found liable for acts or negligence or lack of action on the part of its governing body. When the water utility governing body takes official action which results in the injury to persons or property, legal action can result in the determination of liability and the rendering of a judgment against the utility. In most cases the judgment involves a financial award which must be paid by the water utility to the injured party. There are times when dangerous physical conditions of facilities or faulty operations have been brought to the attention of the officials of a water utility and corrective action has not been taken in a timely way. In such situations the utility can be found liable and subject to penalties in a court of law. The utility manager must continually and routinely be alert to the need for protection of the members of the governing body against legal action as the result of overt acts or negligence on the part of either the governing body or the utility employees.

The types of action or negligence which can result in legal liability cases are almost too numerous to mention. The vehicular accident in which a water utility employee is involved; the clothes which are ruined while being washed by a housewife because water mains were flushed shortly before she started her weekly laundry; the accidental injury to a child who rode his bicycle into a water main repair excavation; the claim by a sick resident that a chemical treatment maladjustment at the water treatment plant caused severe stomach disorder; and the loss of buisness by a businessman because a water main project prevented access to his business property are all real legal liability cases which have been experienced many times. Preventing accidents and other actionable incidents is best, but personal and utility protection against serious or even devastating judgments by means of sound legal services and liability insurance is an essential administrative responsibility to be met by the water utility manager.

When the water utility manager and governing body are planning construction of facilities which will require the issuance of revenue bonds, it is essential that the manager ensure that the services of a bond counsel are retained by the water utility. The bond counsel should be expected to provide advice on all matters pertaining to the sale of bonds and the appropriate bond resolution and other legal documents. The bond counsel should also be relied on to guide the entire bond issue procedure from the time when the bond issue question is prepared for a vote either by the public or by the governing body, up to the time when the bonds have been sold and the necessary procedures have been formally adopted for long-term payment. It cannot be emphasized too strongly that the selection of bond counsel is of extreme importance not only for providing adequate legal advice and ensuring legality of all aspects of the bond issue, but also for the purpose of ensuring that the bonds will have the highest possible market rating and lowest possible interest rate.

Most public water utilities provide services to customers outside the formally recognized water utility jurisdiction, particularly in the case of those water utilities which are a part of a city government. In all such cases service contracts must be negotiated with the out-of-jurisdiction customers after such contracts have been prepared with the advice and assistance of legal counsel. Service contracts with individuals or with other public agencies should provide for all foreseeable contingencies and expectations to the satisfaction of both parties, and should also include provisions which allow certain flexibility so that unforeseen situations will not cause undue hardship on either party. Such flexibility should provide that unanticipated conditions can result in revisions to the service contract, which revisions would be negotiated or otherwise legally settled to the maximum possible degree of satisfaction of all parties to the contract. Inasmuch as most water service contracts remain in force for many years, it is to be expected that changed conditions may require a change in the contractual obligations of the parties to the contract. The service contract should provide for the negotiation of revisions so that the contract will obligate both parties adequately under the law while at the same time allowing the opportunity to relieve an obvious impossible situation for either party. Flexibility of the contract could be achieved by specific provisions for negotiation or for negotiation followed by ultimate arbitration or judicial determination, if necessary, or by any other provisions which would permit corrective contractual action when the need arises.

All contracts involve additional legal concerns for the water utility manager, whether they be construction contracts, professional services contracts, purchase contracts, maintenance contracts, or any other contractual activities. Chapter 13 describes the administrative responsibilities of the water utility manager with respect to contracts in general, but each contract will require full understanding and consideration of all its particular legal aspects and will require the services of legal counsel during preparation of documents, bid opening and evaluation, contract performance, and closing out of the contract. In the preparation and execution of contracts it is important for the utility manager to keep in mind that contracts which require a minimum of interpretation will not only avoid financial penalties, but can avoid the waste of many hours and perhaps days of valuable personal time.

Legal Concerns of the Wastewater Utility Manager

The manager of a wastewater utility is faced with largely the same legal concerns as the manager of a public water utility manager, but the nature of wastewater and the products of wastewater treatment can cause some special legal problems for the wastewater utility manager. (The reverse is obviously also true, i.e., the water utility manager also has unique legal problems.) In considering federal and state laws and regulations which can be troublesome from a legal standpoint, the wastewater utility manager is confronted with a multitude of regulations with which he must comply and which can cause many legal, financial, public relations, and technological problems.

As a case in point, it is probable that the implementing regulations which Congress expected the U.S. Environmental Protection Agency to promulgate in order to implement the Federal Water Pollution Control Act Amendments of 1972 constituted the largest group of implementation regulations ever required of a federal agency in the history of this nation. Many of these regulations constrain the wastewater utility manager as a part of the federal construction grant program, many other regulations provide restrictions and procedural constraints under the planning provisions of the Act, and still other regulations were required for implementing the provisions of the National Pollutant Discharge Elimination System portion of the Federal Water Pollution Control Act.

These voluminous regulations, which have been in some instances promulgated out of phase and in conflict with other regulations, have made it necessary for the wastewater utility manager to make decisions based on less than the best information available, to choose which of two or more conflicting regulations he should or is compelled to violate, and to spend inordinate amounts of time and money seeking legal interpretations of the proposed and final promulgated regulations.

Many of the legal concerns of the wastewater utility manager which are caused by operations and maintenance problems as well as by the planning, design and construction of facilities are similar to the legal concerns resulting from these types of activities in a water utility. Because of the nature of wastewater, both the wastewater, while being transported or treated, and the sludge which results from treatment will at times be putrescent, malodorous, and the cause of various alleged nuisance claims. These troublesome properties of wastewater mean that the legal concerns of the wastewater utility manager are considerably more serious and numerous than is usually the case for water utility managers, whose legal liability in water-related problems normally involves only the damages caused by the water itself, rather than the contents of the water.

In considering the potential liability of a wastewater utility as the result of acts or negligence on the part of employees or the governing body, the wastewater utility manager has the same legal concerns as the manager of a water utility. Likewise, in bond issue matters the legal concerns of the manager of a wastewater utility parallel the legal concerns of a water utility manager. Because a significant percentage of construction costs have for a number of years been borne by federal or state construction grant programs, the dollar amount of wastewater utility bond issues may have been less in most locations than those of water utilities, but the legal concerns regarding bond issues for the two types of utilities are similar.

One type of legal problem which is much more of a problem for the wastewater utility manager than for the manager of a water utility is the nuisance claim. As discussed previously, because of the nature of wastewater it is understandable that citizens will have a tendency to take legal action against wastewater utilities, claiming that odors, pollution of surface water or ground water, fly and insect problems, rodent problems, and many types of illness were in some way caused by the wastewater facilities, the effluent from treatment facilities, or the sludge from treatment facilities. On the other hand, it is obvious that inasmuch as wastewater is not presumed to be colorless and virtually pure, as is water from a public water supply, the wastewater utility manager is not faced with the concern about legal action taken against his utility because of the destruction of or damage to clothes caused by rust-laden water, the loss of tropical fish because of unusually high concentrations of chlorine compounds or other compounds in the public water supply, and many other similar types of water-related problems. In general, however, the manager of a wastewater utility must be more alert to the possibility of the nuisance type of legal action than the manager of a water utility.

Service contracts between wastewater utilities and individual connectors or between the wastewater utilities and other public wastewater agencies can result in legal concerns to the wastewater utility managers similar to the legal concerns or problems resulting from the same types of connectors and water utilities. The legal problems which can result from inflexibly structured wastewater service contracts will correspond to those experienced by water utility managers whose service contracts do not provide for reasonable flexibility, and it is important that the wastewater utility manager ensure as best he can that service contracts with individuals and agencies cover all situations which can be reasonably foreseen, and also include specific provisions for amending the service contract at a later date when found necessary.

Water Rights Constraints

The managers of water and wastewater utilities in most western states, as well as some other states, need to know and understand the implications of water laws and water rights within their particular states. Even though most water law is extremely complicated and more than a water or wastewater utility manager should be expected to understand completely, it is essential that managers have enough of a working knowledge of the applicable water laws to be alert to potential problems in connection with their many projects and activities.

Because of the normal scarcity of and competition for water in arid or semi-arid western states, it is essential that water and wastewater utility managers have access to the advice, guidance, and services of a water rights attorney. Such legal advice from an attorney who is experienced in water rights and water laws can be invaluable when the water or wastewater utility is about to embark on a construction program, when additional sources of water are required and are being investigated or planned, when wastewaters are to be intercepted from an urban area and transported to and discharged at a location which is a considerable distance away, when the water or wastewater utility manager determines that a water reuse or recycling program should be put into effect, and at any other time when a question of the legal rights of other water users may be involved. Legal advice in such cases should be secured from attorneys experienced in the water laws of the particular state and region in which the utility is located and who also are familiar with the political and sociological attitudes concerning water rights in the area served by the utility.

The usual water rights problems which confront water and wastewater utility managers involve the doctrine that the water rights of specific water users may not under normal circumstances be infringed upon by any users who have water rights which are junior to the rights of the other users. This water rights doctrine usually also provides that after a water user has beneficially used his legally apportioned amount of water he may not retain the water or divert it to some other drainage basin, but must release the water for successive use downstream by rightful users. If for any reason an individual water user wishes to make use of water to which he does not have a legal right, he may attempt to exchange water to which he has a legal right with some other water user who has a legal right to that water which the first-mentioned water user wishes to claim for his use. The water exchange concept is important for water and wastewater utilities because under certain conditions the wastewater utility manager may be permitted to provide for the release of effluent from wastewater treatment plants at locations most convenient and most economical for his utility, and it similarly is important that the water utility manager may be permitted to obtain the water supply source for his utility at a location which is most conveniently and economically available for the treatment of such water in the water utility treatment facilities.

The successive reuse of water by a public water utility is desirable from an economic standpoint in most locations where a shortage of water exists or may occur. In these situations prevailing water law may provide that those quantities of water for which the water utility has acquired legal rights and which have been imported from some other drainage basin may be used by the water utility more than once. After final use by the water utility the water would then be released to a stream for subsequent use by some appropriate water user downstream. As the potable reuse of water by public water utilities becomes more desirable and acceptable to the general public and regulatory agencies, it will be essential that the managers of water utilities and wastewater utilities plan jointly to ensure that the wastewater treatment plant effluent is made available for subsequent reuse by the water utility as economically and with as little loss of water as possible.

The proper respect for water laws and water rights will make it possible for the managers of water and wastewater utilities to ensure proper decisions by the governing body in determining how water exchange and other concepts can benefit both of the utilities as well as the people they serve. Other decisions which must be based on sound water rights determinations include water reuse possibilities as opposed to the securing of new raw water sources for the water utility and also decisions concerning possibly higher costs of treatment of wastewater because of stringent stream standards, effluent standards, and the mandatory wastewater treatment requirements of federal or state government. These water exchange, water reuse, raw water purchase, and wastewater treatment decisions can be made intelligently and correctly only at such time as the water and wastewater utility governing bodies and managers have before them the best possible legal advice available concerning water law and water rights.

Summary

The legal concerns of a water or wastewater utility manager will go far beyond the situations cited in this chapter. The intent of this chapter is to convey to those who will be responsible for the management of a water or wastewater utility a message which emphasizes the importance of giving adequate consideration to the legal aspects of all activities, programs, and projects of the utility. It is essential that these utility managers have available on staff or in a consulting capacity adequate and reliable legal advice.

The many liability problems to be found in contract administration, personal injury and property damage situations, nuisance cases, water rights conflicts, bond issue problems and procedures, and many other legal confrontations make it mandatory that suitable legal advice be available to the water or wastewater utility manager on a continuing and reliable basis.

Each overt act of the water or wastewater utility manager, employees, and governing bodies must be taken only after proper consideration is given to its legal implications. In addition, suitable liability insurance coverage must be provided for the protection of the officials and employees of the utility.

10

Public Relations

One of the most important and often one of the most demanding and time-consuming responsibilities of a manager of a water or wastewater utility is the broad area of public relations. If the manager can include in his budget funds for a full-time, or even part-time, public information or public relations staff person, he is fortunate. Regardless of whether the public relations or public information activities of the utility are assigned to a staff person or to a consultant under contract, it is important for the utility manager to realize that the ultimate responsibility for effective public relations for the water or wastewater utility lies with him.

For a public relations program to be effective and significantly beneficial to the utility, it must be well planned and well balanced. The program should include a concerted effort to inform citizens of the activities and programs of the water or wastewater utility and to maintain good communications with those citizens; a routine and continuing public relations effort which will include the indoctrination of employees in public relations attitudes, needs, and procedures and an ongoing effort to project a good image to the public through the utility personnel, activities, facilities, and equipment; a continuing awareness and determination of whether a high level of public information and public relations activity or a low level of such activity is best for the utility; the development of cooperative working relationships with other agencies and associates, including governmental, special interest, and general public groups; and the development and utilization of effective public relations activities throughout the utility organization.

Public Support of Programs and Projects

Most water and wastewater utility managers are fully aware that meaningful involvement of the public in most major decisions of the utility is an important part of the utility manager's administrative responsibility. Not only is the involvement of the public in the decision-making process de-

145

sirable and beneficial, but when a wastewater utility is involved with construction projects certain public participation is mandated by federal law if federal grants are to be utilized for financing part of the facilities construction. There are many effective ways to involve the public in the activities and decisions of the water or wastewater utility, and it is essential that the utility manager attempt on a continuing basis to develop improved and more meaningful public involvement in the decision-making process.

The primary purpose of a water or wastewater utility is to serve the public with safe, adequate, reliable, and economical water or wastewater service. As a part of his determination of how to best serve the public, the water or wastewater utility manager should not only decide what the public needs, but also what the customers of the utility actually desire from their water or wastewater utility. If it appears that the majority of customers desire a high-quality drinking water, it is important that the water utility manager use that knowledge to not only provide the high level of service, but also to take steps to make the public understand that the higher quality of drinking water they desire will result in higher costs to them. Likewise, in the case of a wastewater utility, it is important for the utility manager to determine what level of wastewater treatment the public is willing to finance. The public may wish to only pay for that level of treatment and service which will protect existing water uses, or the level mandated by federal and state law. On the other hand, they may be willing to pay the price of raising the water quality in streams high enough to permit swimming and other water sports, game fishing, or domestic water supply uses.

It is generally understood and accepted that when the public is not adequately informed about new programs, projects, or expenditure proposals, they will normally react negatively to them and will also be less receptive to other changes than will an adequately informed public. It obviously follows that the utility manager must keep his customers reasonably well informed concerning the affairs of the utility so they can react positively to proposed budget increases, recommended bond issues, and any of the many changes which may be needed in the programs and activities of the utility.

The usual reactions of most people to the unknown are hesitancy, skepticism, and even fear. These reactions logically lead to a negative attitude toward the unknown. This is particularly true with the customers of water and wastewater utilities who routinely provide the revenues needed for paying the bills for water and wastewater service. Whenever they are requested to vote for bond issues for facilities with which they are not familiar and whose value they cannot understand or appreciate, they react negatively. This is but one example of an informed public being essential to a program of utility activity.

In some communities it is necessary not only for general obligation bond issues, but also for revenue bond issues to be approved by a vote of the citizens of the community or of the water or wastewater utility. In these communities, it is essential that an effective public information and public relations program be established and maintained so that voter support and approval can be assured when such support and approval are necessary for bond issues.

Even though some people prefer to look at the negative aspects of every problem or endeavor, and as a result develop predominantly negative feeling toward the utility, especially in the case of serious problems, the general public can be expected to support the day-to-day activities of a water or wastewater utility and to provide considerable assistance and support to the utility manager if the manager has conscientiously kept them informed about the utility's various activities, problems, and successes.

The importance of public participation was recognized by Congress, possibly with too much emphasis and concern, in the Clean Water Act of 1977. In that legislation Congress mandated considerably increased public participation in the planning of wastewater facilities construction projects which would be funded partially with federal construction grant funds. It appears that public participation extending far beyond the public hearings as known in the past will be required not only for construction projects to which the federal government contributes funds, but for virtually all activities of water and wastewater utilities. The prudent utility manager will select those public participation requirements which prove to be effective and valuable, and will apply them to other projects and programs of the water or wastewater utility even though such public participation activities are not mandated by federal law. Those public participation requirements which obviously result in no benefits to the general public, to customers of the water or wastewater utility, or to the utility itself should be resisted by both administrative and legislative means.

At those times when new or revised legislation is needed for maintaining or improving the service provided to the public by a water or wastewater utility, it is likely that support and influence from the general public for passage of the legislation can be mustered much more easily if the public has been properly educated by the utility in a continuing public relations and public information program. Legislators on the federal, state, or local level will usually react positively to large members of individuals and groups of people who are able and willing to expend the time and effort necessary to communicate the need for desired legislation. This type of citizen communication can therefore be extremely effective in guiding and influencing legislators, so it is essential that the managers of water and wastewater utilities develop public support for needed legislation.

As mentioned at the beginning of this chapter, one of the most important responsibilities of a water or wastewater utility manager is in the broad area of public relations. To meet these public relations responsibilities, especially with respect to developing public support for the many activities of the utility, the manager must provide adequate, reliable, and understandable information to the general public, service groups, schools, churches, and other public and private organizations within the community. The private citizen, in one way or another, pays the costs of construction, administration, operation, and maintenance of utility facilities and programs, and he thus certainly deserves to have enough understandable information to satisfy his concerns and also to prepare him to fulfill his responsibility as a voter, a lobbyist, and a bill-paying citizen. This information can be provided to him in certain forms as a private citizen, and in various other forms as a member of a church or social group, as an employee or owner of a private business, as the parent of a school-age child, and as a member of a service club or environmental group. An effective public relations program should be so designed and carried out that each citizen will have access to important information.

Day-to-Day Public Relations

Many people think of public relations as being news articles for the various media, brochures with specific purposes, lengthy reports, and speeches. Contrary to this concept, some of the most important public relations efforts and effects of a water or wastewater utility probably involve the daily contact by the public with employees, facilities, and equipment of the utility.

If a public relations program of a water or wastewater utility is to be successful, it is essential that employees who will be in contact with or will be observed by the public receive appropriate indoctrination on and develop an understanding of the importance of daily contact between utility employees and the general public. Such an indoctrination or education program for employees will require some significant expenditure of funds and will require the expenditure of some personnel time during normal work hours, but the results will be worth the effort and expense.

To many citizens the water or wastewater utility is not only represented by but actually is the employee who is seen working in a trench, driving a utility vehicle on a street, or standing in a group studying a problem with a fire hydrant, valve, manhole, or other part of the water or wastewater system. Even though it is unreasonable to expect all utility employees to look clean and neat at all times, a conscientious effort should be required of employees to appear at least reasonably clean and neat in appearance, especially if they are required to be in close contact with or to communicate with people during the conduct of their work. Many utility managers have found it helpful to furnish uniforms to certain or all of their employees to ensure a reasonably clean and uniform appearance. It is at least equally important for employees to be busy and work productively during the work day and for this view of being busy to be projected to the public. At those times when work must be stopped for any reason, whether to collectively study a problem which has arisen, for a coffee break, for lunch, or for any other reason, employees should be oriented to avoid the appearance of idle congregating at a job site. If employees are continually alert to the view had by the public of utility personnel they can avoid creating the impression that they are standing around doing nothing.

Probably as important as the visual impression created by employees on the job is the telephone contact of citizens with utility employees, which may be the only direct contact citizens may have with a public utility. It is essential that all employees of the utility who use the telephone on official business do so in accordance with established high standards of courtesy, cooperative attitude, and a projected impression of reasonable knowledge of the utility and their particular jobs. Persons responsible for transferring incoming calls to officials and employees of a water or wastewater utility should be indoctrinated well enough to know which individuals in the organization should receive certain types of calls. When there is a question as to who should receive a specific call, or if a call is directed to the wrong person, the calling citizen should be asked for his name, telephone number and details of his question or complaint. He should be told that the appropriate person will be calling him within the earliest feasible time frame. The follow-up telephone call by a person who is then at least somewhat familiar with the situation can save the calling citizen the time and frustration of placing several telephone calls and repeating his concern several times to several different persons. The citizen also can be immediately provided with pertinent information at the time of the return call because the employee who is returning the call will have had the opportunity to check into the concern or complaint. This procedure will provide the inquiring citizen with the feeling that the water or wastewater utility is in the hands of employees who care about citizen concerns, know what they are doing, and are part of a businesslike organization. This procedure also should reduce significantly the number of citizens who have a negative attitude toward the utility because of frustration over improper service.

The physical appearance of the vehicles and equipment and the appearance of the employees of a water or wastewater utility can provide important public relations benefits or on the other hand can cause adverse public relations problems for the manager of the utility. Just as employees should be as neat as conditions of their work can permit, the vehicles and equipment of the utility also should present a pleasing appearance to the public. Vehicles and equipment, as much as feasible, should be painted a standard color, should have an identifying decal or other identifying feature, and should be kept as clean as conditions of use permit. When equipment has the appearance of needing a new paint job, it should be painted. A side benefit of keeping vehicles clean and well painted is the psychological lift given to employees, with resulting extra efforts to provide good work to accompany the attractive appearance of vehicles, equipment, and employees.

The public image of a water or wastewater utility can suffer considerably or can be enhanced by the way in which motorists and pedestrains are allowed to move or are forced to move around and adjacent to construction sites and other work areas. Providing for adequate safety and convenience of motorists and pedestrians in such work areas costs only slightly more than exerting the minimum effort, but if adequate safety and convenience are to be provided for the public, the utility manager must insist on planning such provisions into each construction and maintenance job. The unwary motorist should be advised well in advance of arriving at the construction or maintenance site that dangerous conditions are present, that construction or maintenance work is in progress, that it will be necessary for the motorist to move from one lane to the other, that he should watch for and follow the instructions of flagmen or that he should follow detour signs. The providing of the proper types and numbers of signs at construction or maintenance sties cannot be emphasized too strongly, both from the standpoint of safety and from the standpoint of moving the motorist conveniently and rapidly past or around the construction or maintenance site. Detour signs which are inadequate in number or which are misleading are particularly irritating to motorists, and a few bad detour experiences at construction or maintenance sites can make long-term enemies of citizens who otherwise could be helpful friends of the utility. The safety and convenience of pedestrians in their movement past or around construction and maintenance sites is as important as the safety and convenience of motorists. It is particularly important to ensure that pedestrians are not subjected to safety hazards as they walk past such sites. Some of the hazards to be avoided include trenches or other excavations, holes or obstacles within the walkway which could cause citizens to trip or fall, debris which could fall from levels above the street, and possible mechanical equipment, electrical equipment, or water hazards which could cause injury to citizens. Utility workers should be particularly concerned about the safety of children who may not be able to perceive the danger which lies in excavations, running water, construction and maintenance equipment, and other features of construction or maintenance work.

A related type of utility work activity which can often result in adverse public relations with harmful effects on the water or wastewater utility is the daily delay or disruption of traffic by a utility work crew during morning or evening peak traffic movement. Most water mains and sanitary sewers are located within street roadways, and the repair, cleaning or other maintenance of these facilities obviously requires work to be accomplished within the street roadway. The indiscriminate disruption of traffic is bad under all circumstances, especially during periods of heavy traffic movement, and a positive program should be developed which will require that at all times of working within street roadways and street rights-of-way, work crews will maintain traffic movement at the maximum feasible level. Delaying the start of the day's work until after the morning peak traffic period has passed is one requirement which should be considered for work crews even though doing so will shorten the work day. A similar stopping of work within roadways before the evening peak traffic movement occurs should be similarly required to reduce traffic delay problems. There obviously will be situations in which it is necessary to disrupt and delay traffic, especially when shortening the work day or doing the work at night is not feasible. During these times it is necessary that the disruption be reduced as much as possible and that the disruptive work be completed as soon as possible. In cases such as this it is imperative that motorists be advised of traffic delay problems and suggested alternate routes. Such advisory information should be provided through the news media and with signs located well in advance of the immediate approaches to the work area.

If the many construction and maintenance projects of water and wastewater utilities are to be accomplished without serious adverse public relations impacts, it is essential that the utility manager make provisions for a routine motorist advisory system. Such a system should provide signs, barricades, traffic cones, and flagmen which will enable motorists to move from one lane to another as required by the work location with a minimum of delay or stopping, will provide them with the advance opportunity to exit to a different street and drive an alternate route, and will provide motorists with the maximum number of travel lanes and a minimum of conflict with workers or equipment.

One of the easiest and most effective ways by which employees of water and wastewater utilities can establish friendly relations between citizens and the utility is by reacting rapidly and positively to complaints, questions, and requests for service. After a citizen has telephoned or otherwise communicated to utility personnel his concern about something he has experienced or heard, it is essential that the complaint or concern be investigated at the earliest feasible time and the results of the investigation reported to the concerned citizen. If corrective action is required, the citizen should be advised of the probable scheduling of the action along with an explanation of any financial or legal constraints. At the same time the personal support of the concerned citizen for activities of the utility should be solicited. If the utility has no responsibility for or capability of initiating needed corrective action, the citizen should receive advice on whom he should contact, if known, or how he can proceed further with his inquiry. If the concern of the citizen has no basis in fact, some reasonable explanation to that effect should be given to him. The main message which should be transmitted to the concerned citizen is that the utility management understands and sympathizes with the citizen and will do whatever is appropriate in responding to the expressed concern of the citizen.

High- or Low-Profile Public Relations

It is often difficult for water and wastewater utility managers and their governing bodies to decide what should be the appropriate level of public relations and public information activities in which the water or wastewater utility should be involved. If the utility management establishes and maintains a high level of public information activity it is possible that faster and more intense adverse reaction from the public will be experienced at times of failures, accidents, and other serious or incidental problems experienced by the utility than would be true if the utility has a low level public information program. On the other hand, it is reasonable to believe that maintaining of a high public information profile in the community will enable the water or wastewater utility manager to more easily appeal to and convince the public of the necessity for supporting a bond issue, a legislative program or effort, a specific program or project or the legal position of the utility when such public support is needed.

An important part of any effective public relations program is a well-designed and well-executed public attitude survey, which should normally be conducted by professional public information or public relations personnel who are experienced in the design and conduct of such surveys. The public attitude survey will indicate the level of awareness among the general public and within specific segments of the community about the water or wastewater utility, its programs, its successes, its costs, and its problems. The public attitude survey can provide the utility manager with a reasonable indication of the understanding by the public of the general mission, costs, and problems of the utility as well as the need for facilities construction, rate increases, or legislative action. The results of the public attitude survey should enable the utility manager to determine how receptive and responsive the public may be to intensive or subdued public relations activities and the type of information which can be expected to provide the desired understanding and effect. Additional followup surveys can indicate the effectiveness of the public relations efforts.

An evaluation of the probable effects of a high level of public relations activities as opposed to low level public relations activities requires an analysis of the probable and possible advantages and disadvantages of the two levels of activity and the impact of each on the many activities of the water or wastewater utility. An evaluation of the probable impact of a public relations effort of moderate profile or level also would be appropriate. The advantages of high-level public relations and public information programs normally would include an increased awareness on the part of the public of the needs of the particular utility when support is needed for a bond issue for construction of facilities or for legislative action in either a state legislature or in the United States Congress, and in similar circumstances. Other types of support for various activities of the water or wastewater utility can probably be enhanced with a high-level public relations program which would tend to make influential citizens and politically motivated groups or individuals not only aware of the activities of the utility, but also sympathetic toward and supportive of the specific needs of the utility management and governing body. Such support can influence news media to report in a positive rather than in a negative manner on successes and failures of the utility or on budgetary needs of a routine or special nature, and to provide positive rather than negative influence on various regulatory and permit-issuing agencies with whom the utility manager must contend. An obvious disadvantage of high-level public relations or public information programs is the emphasis which may be placed on any failures or mistakes of the utility, however insignificant they may be. In some situations public notification of failures obviously would be detrimental to the utility, and the manager would prefer that the failue would not be publicized out of proportion to its real significance.

Some of the advantages of a low public relations profile or of low-level public information programs include the avoidance of or at least a reduction in the amount of unfavorable publicity which can accompany insignificant, as well as significant, failures or problems of the water or wastewater utility, the reduction of opportunities for news media to sensationalize utility activities in a negative manner, and also the elimination of those situations in which the utility can become a political target for politically minded individuals or groups. For those utilities which maintain a low public relations profile, it is still relatively easy for important news stories which can be helpful to the utility to be covered positively and adequately by the news media if a friendly and cooperative relationship with news media personnel has been maintained in the past. In addition, with a low-level public relations program it is still not difficult to enlist the support of individuals, special interest groups, and various associations and agencies for the purpose of supporting important bond issues, legislative efforts,and other important programs, projects, and activities of the utility.

It is often advantageous for a water or wastewater utility manager to employ a public information or public relations program which is alternately high-profile and low-profile in nature. In utilizing alternating high and low profiles the utility manager can concentrate on public relations and public information activities at times of bond issue campaigns, when seeking approval of the budget by the governing body of the utility, when seeking legislative support of important programs, and at other times when public involvement is needed. During times of relatively routine activity the utility manager can retain a low profile for the utility and still provide newsworthy information to the media, both to maintain good relations with the media and to keep the general public informed on those matters which the utility manager and governing body consider important.

Whether the public relations-public information efforts of a water or wastewater utility are high profile, low profile, or alternating between high and low profile depends on many factors. The relationship of the utility manager and other utility officials with reporters is a major factor, especially in cases of extreme cooperation and friendliness or extreme antagonism. The attitude of the utility governing body toward public information programs certainly will determine the type of public information programs to be followed by the utility manager. The quality of public relations and public information talent on staff can be expected to be a limiting factor. Other factors can include the fiscal stability of the utility, its operational and administrative successes or failures, the need for bond issue financing, the relationship of local utility rates to the utility rates of other agencies, and many other events and attitudes within as well as outside of the influence of the utility manager.

It is important for the manager of a water or wastewater utility to fashion the public relations program to the real-life situations which exist and which are anticipated. The manager must retain flexibility in the public relations program so that positive attitudes can be developed among citizens toward the utility at those times when activities of the utility result in recognizable progress and success. The flexible public relations program likewise will enable the utility manager to minimize the negative impact on public attitude toward the utility as the result of minimizing any reporting of failures.

Relations with Other Agencies

In the routine management of the many activities of a water or wastewater utility, the manager must continually interact with officials and personnel of many other agencies and organizations, including agencies of the various levels of government, special interest groups, public interest groups, and of course, the news media. During the course of these continuing interactions the utility manager must maintain a continuing public relations program with all levels of government, including federal, state, county, regional, and city governmental agencies. As personnel changes occur within the various governmental agencies it is essential that the utility manager establish the necessary new contacts with the replacement personnel so as to minimize disruptions in the established contacts and relationships with the specific agency and any of the individual personnel of the agency. Continuity of working relationships with the appropriate personnel of various governmental agencies is essential to a smooth working relationship between the utility and the agencies. Individual working relationships must be developed and maintained with those agency personnel who are specifically responsible for the enforcement of laws and regulations, construction grants, research and development, safety, civil rights, and education and training.

The working relationship with the various news media is of prime importance to the water or wastewater utility manager if the general public is to be expected to receive factual information on utility matters, and if the public is to be influenced to play a supportive role rather than an opposition role concerning the activities of the utility. The release of factual information to news media on a regular basis is essential to the development and maintenance of good news media relations. Even when the utility manager may feel that he has nothing newsworthy to release to the media, he should be easily available to reporters to answer questions and discuss current matters with them. Through this process the utility manager can provide news media personnel with the opportunity to obtain information which might lead to stories which could be important from their viewpoint. As news media personnel or personnel assignments change, it is essential that continuity of established relationships between the media and the utility manager be maintained. The utility manager should attempt to develop such a sound working relationship with reporters that they will forewarn the utility manager of contemplated personnel changes so as to avoid discontinuity in the news media relationship. It is important that the utility manager avoid the situation in which he contacts reporters only when he wants help or favors from them, rather than routinely offering information and prospective stories. It is much easier for the utility manager to request and expect to receive special coverage of a specific utility matter if he has conscientiously made it a practice to routinely provide newsworthy information to the media.

Because of the need for the utility manager to solicit and receive supportive help from environmental groups, school officials and students, voters, and service clubs, it is necessary that he establish friendly, trustworthy, and helpful working relationships with such groups.

Environmental groups are the source of a great deal of important support for many of the programs of water and wastewater utilities, but such groups also can cause considerable delay and additional expense if they oppose construction projects and certain other activities. The development of a positive, friendly and cooperative relationship which provides for free exchange of information, opinions, and ideas between the utility manager and the members of individual environmental groups is an important part of the successful management of a water or wastewater utility.

The administrators and teachers of the various public and private schools in the community should be sought out to develop a continuing program of tours and educational programs for their school students offered by the personnel of the water or wastewater utility. These efforts obviously should be accompanied by a soliciting of support for activities and programs of the utility from school administrators and teacher groups. Many teachers of university, secondary, and elementary students are anxious to include tours of facilities such as water or wastewater treatment plants as a part of their course curriculum, especially for science, biology, environmental, governmental, and similar courses. The opportunity to develop a mutually supportive program with teachers and school administrators as well as with students should be enthusiastically grasped by the utility manager. To the wise utility manager it should be obvious that the influence of schoolchildren on their parents can be an effective method of reaching the adult citizens in the community for support of any of the programs, projects, or other activities of the utility.

Voter groups such as the League of Women Voters are dedicated to the dissemination of factual information to voters and the total improvement of voter understanding of election procedures and election positions. The utility manager should work closely with local voter group leaders to develop a meaningful interchange of information and ideas as well as to provide to the utility a valuable source of guidance and counseling in matters pertaining to elections, legislative influence, and voter attitudes. Even though such groups may not take an official position on voting issues of concern to the utility manager, their advice and dissemination of information to the public can be of great value. The manager of a water or wastewater utility should develop a working relationship with service clubs in a community similar to the relationship which should be developed with environmental, school, and voter groups. Service clubs are normally composed of service-oriented people in the community, and are usually anxious to be of assistance to local officials in both their routine programs and their special programs of serving the public. Service club program committee chairmen are usually looking for interesting and meaningful programs for their clubs, and the presentation of programs to service clubs by the utility manager or his appropriate representatives should be a routine part of the public relations activities of the utility.

One particularly effective way in which the utility manager can assure a satisfactory relationship with local public and private groups whose assistance he may require for bond issue support or any other reason is to participate personally in the activities of those groups. This requires that the utility manager be personally, or through a representative, active in the many activities of school, church, service, environmental, and voter groups as well as in other organizations if participation can result in mutual benefit. By their active participation the utility manger and his personnel can more easily influence the groups to be supportive of the activities of the water or wastewater utility.

Participation and cooperation by utility managers and their personnel in the activities, especially the problem-solving activities, of other water and wastewater utilities provides the opportunity for an interchange of valuable information and ideas which may not be possible or at least may not be as effective without such participation and cooperation. Active participation in associations of water and wastewater utilities also provides the opportunity to exchange ideas and to seek assistance from experts in the field when assistance or information is required. This sort of participation by a utility manager and his employees obviously requires a considerable expenditure of personal time if it is to be of value, but the benefits are certainly worth the time and effort. (It is necessary, however, to exercise discretion and sound judgment in this regard to ensure that participation is not excessive to the point of neglecting other responsibilities.) The utility manager has a special responsibility to advise and convince the governing body of the need for and value of such participation in the affairs of professional and technical associations.

An effective public relations activity which can be valuable to a water or wastewater utility is in the form of a citizen advisory committee or any of several other types of advisory committees. Such advisory committees should be established for specific purposes such as bond issue campaigns, facilities planning, consideration of revising water or sewer rates, and organizational changes. Some advisory committees of a continuing, general type have been valuable, but the success of this type of committee will require that the utility manager devote considerable attention to the agendas of the meetings of such committees to avoid boredom and a feeling of wasting time on the part of committee members. It is important to select committee members who are representative of a reasonable cross section of the community, are capable of providing the time and effort required of committee members, are genuinely interested in the activities of the utility, and are not directly involved in the management of the utility. It is usually wise to include on the committee at least one or two people who are negative or at least neutral concerning the utility activities to be considered by the committee. This should help avoid the pitfall of basing the recommendation of the committee on preconceived notions which supporters of activities normally have. The negative attitude provided by some members of the advisory committee can provide valuable views which can be used to revise the activity under consideration or can enable the utility manager to develop an effective informational program to eliminate or reduce the negative attitudes of other individuals in the community whose views are basically identical with the negative stand of some committee members.

Internal Public Relations

The routine communications and interactions of a water or wastewater utility manager with his governing body may appear to not really be a part of the public relations program, but it may be the most important public relations activity in which the utility manager is involved. The utility manager has the basic obligation to manage the total affairs of the utility under the policy guidance of the governing body. The management of the utility should be performed in such a way as to ensure proper operation and maintenance of facilities, adequate technical and fiscal planning and effective control of the full range of administrative activities of the utility organization with a minimum of interference by the governing body. At the same time the manager must involve the governing body to the extent necessary to keep it adequately informed and thus enable it to provide the required policy guidance for the utility manager and the organizational structure. If the utility manager is to maintain an effective relationship with the utility governing body, including a satisfactory exchange of ideas, attitudes, and information, he must be flexible in his relations with the individuals on the governing body. At times it will be possible for the manager to resort strictly to factual information to enable the governing body to make appropriate policy decisions, but at other times it will be necessary for the manger to resort to presentations based on emotions and attitudes rather than factual, technical information. Even though major decisions by the utility governing body should be based on facts and logic, the experienced utility manager will determine the type of presentation which will be most effective with his specific governing body for the specific needed decision and will use that type of presentation. At all times, however, it is imperative that the manager have adequate facts and logic to support his position regardless of the method used to influence the governing body in coming to their ultimate decision.

A somewhat similar aspect of internal public relations which must receive much attention of the water or wastewater utility manager involves the communications and internal relations which the manager establishes and maintains with the personnel in his upper management positions. If the utility is to maintain a record of satisfactory performance, the manager must keep his upper organizational levels well informed on all aspects of the activities of the utility and must involve such personnel, in a meaningful and participative manner, in both routine and special decision-making. It is essential that key personnel within the utility organization understand specifically what is expected of them in the way of performance and decision-making. It is equally important for such key personnel to understand the limits of their authority in making decisions concerning administrative matters and operations or maintenance actions as well as their limitations in their contacts with the various segments of the general public, the governing body, the news media, and various governmental agencies.

One other segment of the total internal public relations responsibilities and efforts of a manager concerns the important communications within the utility organization between the personnel who make up high-level management and the non-management employees of the utility. The broad subject of employee relations will be covered in a later chapter, but it is important in the consideration of the total concept of public relations and public information to emphasize the importance of ensuring that utility employees are kept well enough informed so that they can be an asset to the total public relations and public information activities of the water or wastewater utility. It is a known fact that if employees of an organization feel important to the success of the organization and understand the need for them to assist in the public relations program of the organization, they will be considerably more effective in their contact with the public than they otherwise would be.

A Few Examples of Effective Public Relations

A number of examples of effective public relations have been presented throughout the chapter, but at the risk of repetition a few additional detailed comments will be made concerning particularly effective public relations activities. The techniques described may not be unique or particularly new, but they have been found to be effective and can serve as good examples of many of the various types of public relations activities which can be extremely helpful to the water or wastewater utility manager.

As discussed previously, the impression that many citizens have of a water or wastewater utility, its management, and its personnel can be based entirely on a single telephone inquiry and conversation. Whether the purpose of the telephone call is to voice a complaint about unsatisfactory service, to inquire about how a water or sewer problem in or around the home can be solved, or for any other reason, it is imperative that all employees of the utility who would have reason to speak to inquiring citizens on the telephone receive training and indoctrination in the proper way to handle such calls. Of paramount importance in such indoctrination is emphasis on courtesy and helpful, friendly attitudes. In the event that the receptionist or person who is called by a citizen is not qualified to adequately respond to the inquiry, and if there is a probability that the citizen will be referred to more than one additional person, it should be routine procedure for the person who is receiving the call to request from the caller his name, address and telephone number and the nature of the inquiry or complaint for the purpose of assuring a return call from a utility employee who is qualified to handle the complaint or comment. The caller should then be advised that the appropriate person in the utility organization will call the citizen within a few hours or the next day, depending upon the inquiry and the time needed to respond satisfactorily to the call.

One of the instructions which should be given to all personnel is that under no circumstances is a telephone caller to be switched from one person to another for the purpose of eventually reaching the appropriate employee. In virtually all cases it is best to require an employee of the utility to return the call to the citizen after he has had an opportunity to give at least preliminary thought to the inquiry or complaint of the citizen. Of course, if at the time a call is received the person who receives the call can determine the specific employee who should handle the matter and knows that this employee is available and is prepared to talk to the caller, the telephone call may be switched directly. The important point to be made here is that nobody calling for information or with a complaint should be referred to several people with each employee passing the increasingly impatient and irritated citizen on to somebody else. In addition to antagonizing the caller, this practice will always give the impression that employees of the utility are incompetent and that the utility is disorganized.

The prudent use of construction information signs together with warning signs and safety devices normally available for identifying work areas can provide a multitude of benefits to the water or wastewater utility as well as to the public. A well coordinated and well planned system of signs, barricades, flashers, flagmen and detours is required to provide safe movement of motorists and pedestrians through, past and around the sites of construction and maintenance work, and this safety feature is a particularly important benefit to the public. In addition to providing safe movement of motorists and pedestrians past construction and maintenance sites, the use of appropriate signs can relieve some of the negative or even antagonistic public attitude to construction and maintenance activities, which often cause traffic disruption, detours, and traffic delays. Construction information signs should be provided in a form which is pleasing to the eye, is simple, and includes pertinent and interesting information about the specific project. Such information should include a descriptive title and the purpose of the project, identification of the agency responsible for the project, the project cost, its probable completion date and other descriptive information which could be meaningful to the public. An apology on a construction or maintenance project information sign for the inconvenience caused by the project to the travelling public and the nearby homeowners can be effective in turning inconvenienced and often belligerent citizens who are close to becoming enemies of the utility into understanding citizens, and can even gain friends among them. The information presented on construction project signs can keep the public up to date on the status of specific projects which are funded by bond issues approved by the voters as well as projects which are funded in some other way, and can help to convince citizens that their supportive votes are actually resulting in important construction work.

Signs, valuable as they are, are no more important than barricades and detour systems in these construction and maintenance work areas. Barricades should be used at all sites where it is important to warn driving and walking citizens of dangerous conditions and guide them away from potential dangers. The barricades should be clean, well painted, preferably reflectorized, and equipped with flashers if they are to be in place at night. Appropriate signs should be utilized together with the barricades to provide proper guidance to motorists and pedestrians to ensure their safety and to avoid confusion. In the event that a traffic detour is required at a construction or maintenance site, it is imperative that adequate signs, arrows, and barricades be provided to guide the motorist completely through the detour with no need for guesswork at any intersection. Responsible personnel should be instructed to use additional signs if any doubt exists as to whether more or fewer signs are needed. An important part of effective detour operation involves the retiming of traffic signals at intersections, which should provide for a larger percentage of the signal cycle green time to the detour route than to the cross streets if the detour route traffic volume is greater than the cross streeet traffic volume. This feature is often overlooked in the planning of a detour route and could be the most negative feature of a detour from a public relations standpoint.

If the utility manager is to ensure the involvement of employees in effective public relations he must provide a continuing indoctrination and training of employees in the need for and methods of effectively dealing with the public. New employees who will be in contact with the public by telephone should receive extensive instruction and training in the proper use of the telephone, especially courtesy; information concerning the routine and special activities of the water or wastewater utility; information about the key officials and personnel within the organizational structure; and other pertinent information which they may need in properly handling telephone calls. Those employees who will be seen by the public while working, resting, or operating utility vehicles and equipment whether on the streets or in other locations should receive indoctrination which will enable them to understand the necessity for appearing neat and busy; how to respond to inquiries from citizens while they are on the job; how to continue working on their assigned jobs without offending the citizen who wishes to converse with the employees; and other similar types of information which can enable employees to be aware of how they can make a positive public relations impact on the public.

The vehicles and other equipment of a water or wastewater utility can be used to aid the public relations program of the utility by keeping them in a clean, well painted, and properly marked condition to encourage a positive public attitude toward the utility. The normal citizen will assume that equipment which is dirty or badly in need of a paint job reflects the quality of workmanship on the part of utility employees. The equipment should be identified as being property of the water or wastewater utility with an identification decal or lettering together with a vehicle number so that citizens who have a complaint or inquiry concerning a specific item of equipment can accurately identify it. Identification of equipment is also important so that employees realize that any improper action on their part can be easily reported by a citizen who may see it.

One final example of effective public relations to be considered here relates to uniforms and proper identification for employees, especially those employees who will be in contact with citizens for reading meters, for answering complaints or inquiries, or for advising citizens of some activity of the utility which will have an impact on the citizen. The uniform can be used both to identify the employee as being employed by the utility, and also to bring about an attitude of identification of the employee with the utility and pride in the utility, including the physical appearance of the employee.

These few examples should serve as a guide to the utility manager for developing public relations and public information efforts on the part of his employees which can assist the public relations program of the utility as well as or even better than expensive formalized public information activities. It is important for the water or wastewater utility manager to fully comprehend the impact his employees will have on the total public relations effectiveness of his utility.

Summary

The manager of a water or wastewater utility should devote considerable effort and much of his personal time to the public relations activities of the utility.

Public support of the programs and projects of the utility can be assured with an effective total public relations program, involving virtually all employees of the utility. This public relations program can be planned to be high profile or low profile or a combination of both, taking into account the reactions of the public and the specific needs of the utility. Cooperative interaction with other utilities, agencies, and organizations is an important part of the program, and significant benefit can result therefrom. A strong internal public relations program is also essential.

11

Personnel Planning and Employee Relations

Payroll costs usually account for approximately 50 percent of the total annual operations and maintenance budget of a water or wastewater utility. With such a large portion of the budget devoted to personnel expenditures, it is not only to be expected, but it is imperative that the manager will ensure that adequate consideration and time are given to personnel matters by all levels of management and supervision within the organization, including a significant amount of personal time on the part of the utility manager.

Every aspect of personnel administration should receive nothing less than top priority attention of the utility manager. This includes the determination of staffing needs, the recruitment and development of qualified personnel, the motivation of individual employees, the development and maintaining of adequate communications between management and the employees, the recognition of and participation in the collective bargaining and union relationship activities of the utility, and the administration of the entire wages and fringe benefits program.

Determining Staffing Needs

The evaluation and determination of staffing needs is necessary not only for new or expanded facilities, or for facilities within which the operations or maintenance activities have undergone considerable change, but also on a continuing basis for any water or wastewater facility. The continuing routine analysis of staffing needs for water or wastewater utilities should be in the form of a periodic personnel audit, a procedure whereby the utility manager can be advised, with appropriate documentation, whether additional personnel are needed for specific activities, whether excess personnel are on the payroll and in what classifications, what additional types and levels of skills

163

are required, and whether other types of personnel action are needed to maintain proper staffing.

In determining staffing needs it is essential that the utility manager first determine which specific tasks must be accomplished within the utility organization and then identify the types and numbers of personnel required for each related group of tasks. The identified tasks can then be tentatively assigned either to the various personnel classifications or to computer process control or monitoring equipment rather than to personnel classifications, depending on feasibility and cost effectiveness of each. Inquiries to other water or wastewater utility managers concerning their classifications and numbers of personnel assigned to various functions and activities is one means of evaluating the personnel staffing of a utility organization. Even though the organizational structures of other water or wastewater utilities may be considerably different from that being evaluated, and though the other utilities may be overstaffed or understaffed, considerable benefit can result from studying a number of other organizational staffing patterns with which one's own organization can be compared.

If part or all of anticipated future treatment facility monitoring or process control will be performed to a considerable extent by a computer system, the staffing needs for such a water or wastewater treatment facility will be understandably different than if the monitoring or process control will be accomplished manually. Manual monitoring will require samplers or additional operators to provide the necessary sampling for monitoring purposes, as well as laboratory technicians to perform the appropriate laboratory analyses. The computer system obviously will require personnel for equipment maintenance and calibration as well as for operation of the equipment.

When adopting a procedure for determining staffing needs, it is generally helpful to estimate the staffing requirements by analyzing groups of various processes and activities. Such groupings could include water or wastewater treatment; solids processing and disposal; plant equipment maintenance; maintenance of vehicles and mobile equipment; other facilities and grounds maintenance; laboratory activities; accounting and purchasing activities; and the other administrative, managerial and supervisory responsibilities. In some cases, it will be determined that certain individual personnel will actually be assigned to several of the groups such as to wastewater treatment plus solids processing and disposal or to equipment maintenance plus facilities and grounds maintenance. The physical size of the facilities as well as the number of locations at which similar facilities are located will be a major factor in deciding whether specific employees will be assigned to a single facility, to a single job at a single facility, to a single job at several facilities, or to a number of jobs at more than one facility.

The determination of management and supervision staffing needs normally will be based on the numbers of specific classes of personnel required for the various activities of the water or wastewater utility. In some cases, one supervisor could be expected to provide adequate supervision for as many as ten to fifteen people who have similar skills and are physically located near each other; in other cases a single supervisor, because of the geographical separation of personnel to be supervised or because of the nature of the jobs being performed, may only be able to provide adequate supervision to five or six. Management personnel needs will normally be determined on the basis of the types of management and administrative activities. The activities for which administrative and management personnel would be needed include operations and maintenance, possibly separate solids processing and disposal activities, laboratory analysis and monitoring, engineering, planning, and administrative services such as accounting, purchasing, and personnel administration.

The studying and planning of personnel needs and assignments should be completed or at least near completion during the design phase for new facility construction or facility expansion. When the utility manager has a clear understanding of the additional facilities to be operated and maintained following completion of construction, he should complete the process of determining the personnel of various classifications which will be required for the operation, maintenance, and monitoring of the new facilities together with the personnel required for the many supporting activities in connection with the new facilities. Similar planning of personnel changes obviously is required for the case in which one or more operational changes will necessitate the hiring or training of personnel with skills different than those possessed by present employees. Here again, it is important that the utility manager determine at an early date the change in numbers of various classes of employees so as to ensure adequate cross-training of present employees, hiring of new employees, and, if jobs will be eliminated, helping to find jobs for those employees who will be terminated.

Assistance of considerable value in determining staffing needs for new or existing water or wastewater utility facilities can be received from consulting engineering firms, other similar operating agency managers, and the U.S. Environmental Protection Agency. Even though each of these sources of information must be used with questioning judgment, they all can provide valuable input to the personnel planning of the utility manager. The ultimate determination of staffing needs will result from a combination of these sources of information with the local experience of the utility manager. The local experience should be the major factor in the staffing decisions, especially if work measurement methods have been employed to determine the work performance which can be reasonably expected of the various employee classifications in the various jobs.

Planning for staffing of water or wastewater utility organizations should include a thorough understanding of the labor market in the community with respect to not only skilled personnel and semi-skilled personnel, but also for non-skilled personnel. The availability of supervisory personnel of the various required qualifications as well as clerical and lower-level management and professional personnel must be determined both as to the present condition of the labor market and the probable trend in the near future. If the labor market does not appear to be adequate to meet the staffing needs of a utility organization for new or expanded facilities, or because of operational changes or for other reasons, plans should be made for recruiting the required talent outside of the local area. If recruitment of personnel from outside of the local area appears to be infeasible for some reason, a training program for upgrading personnel of lower-than-needed skills must be instituted. If there will be inadequate numbers of job applicants of even the lowest skill levels available within the community, there will obviously be no alternative to recruitment of personnel from outside the local area.

At the time of determining staffing needs it is wise for the utility manager to plan into his staffing needs as many advancement possibilities as possible for present and prospective employees. This may require that some skilled positions must remain unfilled for a time or filled with unskilled personnel who will require training of various types for advancement into the higher skill-level positions. This will enable the utility manager to provide many of the personnel in his organization with a maximum of advancement possibilities with accompanying positive employee motivation and morale benefits.

In addition to the need to at least commence, and preferably to complete personnel planning at the time of facility design, it would be valuable for the utility manager to provide for a maximum of coordination of facility design and future maintenance needs. Many opportunities will present themselves to incorporate into the design certain features which will minimize maintenance needs, will permit the planning of improved maintenance coverage for isolated areas of the facilities, and will ultimately result in maintenance staffing being set at a minimum. Since the annual payroll cost for a water or wastewater utility is about 50 percent of the annual budget, it is imperative that the utility manager continually look for ways in which personnel can be kept at minimum levels without jeopardizing the quality of operation or maintenance of the facilities.

Examples of Actual Staffing

Each water or wastewater utility has numerous special characteristics which influence the numbers and types of employees in its organization. These characteristics include: either water or wastewater responsibilities versus both water and wastewater; responsibility for water or wastewater treatment facilities plus distribution system or collection system versus only a part of these total systems; labor unions versus no unions; high level of water or wastewater treatment versus lower level of treatment; old water or sewer system versus new system; and many other characteristics.

It is thus essential that staffing needs be determined with as much outside assistance, guidance and advice as possible. To provide some additional guidance in staffing needs determination, information concerning the actual staffing of three utilities is presented here.

Utility A. This water and wastewater utility serves 5,013 water customers and 5,736 sewer customers, and is responsible for a water treatment plant, a water distribution system, a surface water supply, a sewer collection system, and a trickling filter wastewater treatment plant.

Personnel include a utility manager; an assistant manager; four office personnel including an office manager; one utility superintendent; one water treatment plant superintendent; six water treatment plant operators; two wastewater treatment plant operators; one person who works at both treatment plants; and eight maintenance personnel for the water and sewer systems together (60 percent of the manhours for the water system and 40 percent for the sewer system, including maintenance of the treatment plants). Thus for approximately 5,500 water and sewer customers (approximately 14,000 population) Utility A has 6 administrative employees and 19 technical workers.

Utility B. This water and wastewater utility serves approximately 29,000 customers (approximately 80,000 population) and is responsible for water supply from surface sources, a water treatment plant, a water distribution system, a wastewater treatment plant which treats about 20 percent of its wastewater and a sewer collection system (80 percent of the wastewater is treated by a regional wastewater authority).

Personnel include a utility manager; 17 office and administrative personnel; 3 superintendents for wastewater, water treatment, and water distribution; 4 wastewater treatment plant operators; 1 wastewater technician; 10 sewer collection system personnel; 1 water treatment plant mechanic; 10 water treatment plant operators; and 18 personnel for construction and repair of water distribution system facilities. Thus for approximately 29,000 water and sewer customers (approximately 80,000 population), Utility B has 18 administrative and office employees and 47 technical workers.

Utility C. This regional wastewater utility serves approximately 400,000 customers (approximately 1.1 million population) and is responsible for 65 miles of interceptor sewers and an activated sludge wastewater treatment plant, but no sewer collection system.

Personnel include a utility manager; an administrative services department of 25; an engineering department with 8 employees; a laboratory services department of 20; an operations and maintenance department with an operations staff of 32 and a maintenance staff of 57; and a sludge disposal department with 42 employees. Thus for approximately 400,000 sewer customers (approximately 1,100,000 population), Utility C has 35 personnel who could be classified as administrative employees and 149 technical workers.

Inasmuch as these three utilities are of considerably different sizes, serve widely differing numbers of customers, and have very different responsibilities, no effort should be made to compare their staffs. These staffing levels, however, can obviously be of some help to utility managers in determining their own staffing levels.

Establishing Qualifications for Positions

After the staffing needs have been determined for a water or wastewater utility organization, it is necessary to establish the qualifications which will be required for specific positions. These qualifications can be determined by studying the tasks which must be performed by specific employees and identifying the skills which will be needed for those tasks. The necessary skills must then be provided to employees and prospective employees through a combination of education and training, practical experience, aptitude, and motivation. In some job classifications, in addition to skill requirements, certain physical qualifications also will be required of employees. These physical qualifications should be incorporated into the relevant job descriptions.

The establishment of the qualifications required for specific positions must take into account the desirability of staffing not only with employees who are already completely qualified, but also with employees who are relatively inexperienced. To this end, it is desirable that minimum qualifications be used as requirements in job descriptions for entry level laborers, for semi-skilled laborers, for skilled operations and maintenance personnel and laboratory personnel, and in most cases for subprofessional and professional employees. The establishment of qualification levels for the many levels of employee skills will make the path of advancement and required training meaningful for employees who possess relatively few skills.

Wage scales for the various employee classifications should be established in reasonable accordance with the wage scales in the community and surrounding areas. In the event that the prevailing wages for anticipated work categories are not adequate in a given community to attract the needed skills, and the utility governing body is not willing to pay higher than prevailing wages, it will be necessary for the utility manager to determine whether fewer employees can be utilized for the required tasks if personnel with considerably higher skills than previously assumed are employed. If it appears that this strategy will not be feasible, it may be necessary for additional employees to be used at the lower wage scales with less production per employee or fewer tasks per employee to be expected.

The proposed qualifications for each employee classification should be compared with those of other similar agencies, both in the general area and elsewhere in the country, for the purpose of determining whether the qualifications and accompanying wage scales will be adequate for the needs of the water or wastewater utility. As a word of caution, it should be emphasized that blindly accepting the results of such a comparison can be less meaningful than proceeding with the determination of required qualifications and wage scales with no comparison with other agencies or other outside sources of information. This type of information should always be used as only one of the many factors which are considered in the decision-making process.

The required qualifications as well as the desired qualifications for each employee classification should be established at levels which ensure good employees for the utility. However, the qualifications should also be reasonable, especially in a relatively tight job market. It will often be difficult for the water or wastewater utility manager to impose sufficiently stringent job qualifications for the needs of the utility while at the same time ensuring that there are enough qualified people to fill available positions. Again, it should be emphasized that the setting of qualifications for job classifications of several related skill levels should be done so with special consideration of the planned paths of advancement and promotion for as many employees as possible.

For those employees who will be responsible for operating or maintaining equipment of various types, the recommendations of the equipment manufacturers relative to personnel abilities should be solicited, evaluated, and utilized as considered appropriate. For many types of equipment used in the water and wastewater fields, the manufacturer of the equipment is often one of the best judges of the necessary qualifications for operations and maintenance personnel. It is essential that the utility manager use sound judgment in considering the recommendations of manufacturers' representatives together with his experience and the advice of other people in the water or wastewater utility field.

Education and Training Programs

An effective education and training program is an essential function of any successful water or wastewater utility, and this program should be made available to all levels of employees: unskilled labor, employees with trade skills, and the clerical, professional, and administrative personnel of the organization. It is to be expected that after some employees, including some of the more highly qualified employees, become experienced and well trained they will go on to some other higher level position for some other employer at a higher salary. Some may bemoan this fact, but a utility manager should look on these losses as proof of the success of his total education and training program and they should not deter him from encouraging his employees to pursue education and training courses.

A considerable part of the education and training of utility employees can be provided in-house, with personnel of the utility serving as instructors. Outside specialists should, of course, be used as expert lecturers for many of the specific topics in the program. In-house education and training can be both on-the-job training and formal courses conducted either during or outside of work hours. Formal short courses can be provided for many levels of employees in the different specialties, again either provided by personnel within the utility organization or made available at nearby colleges and universities, high schools, and trade schools. These short courses can be financed by the water or wastewater utility or can be under the sponsorship of the utility, but financed by the employee.

An important employee educational benefit which can be valuable to both the employee and the utility is reimbursement of the cost of tuition, books, and laboratory fees for courses taken by employees at nearby universities or colleges. The reimbursement would normally be limited to some amount such as $300 per year, and should be limited to only those courses which have some reasonable relationship to the employee's present and possible future job with the water or wastewater utility. When properly administered, this program can, in addition to providing increased expertise for employees, serve as a motivational factor in improving their work performance.

One of the most important types of education which can and should be provided to certain of the water or wastewater utility employees is in the area of public administration or utilities management. It is essential that utility employees who may have an opportunity to be promoted into management positions within the utility receive formal public administration education to prepare them for the responsibilities which will be theirs when they are promoted into management positions. A considerable number of universities throughout the United States have developed graduate programs in public administration, and a few of these have also developed public works administration courses, public works administration options in the master of public administration degree programs, or specialized public works administration graduate degree programs. Although student participation in some of these graduate programs may be required to be on a full-time basis, most of these courses can be pursued part-time. Courses can usually be taken in evenings, and one or two such courses can be pursued by an employee during each school term while he continues to work full-time. The utility manager will find it necessary to expect a loss of employee work time for some courses, but the ultimate benefits will far surpass the time sacrificed by the utility.

The value of technical and professional conferences and conventions often is not fully understood or appreciated by the general public or the governing body of water or wastewater utilities. It is a responsibility of the utility manager to ensure that key personnel in management, professional, and supervisory positions, as well as certain other personnel, have an opportunity to attend those conferences and conventions which provide meaningful and valuable technical programs, equipment and systems exhibits, and the opportunity to confer with professional consultants, researchers, equipment distributors, and other utility employees from throughout the country and the world. The conference or convention which is predominantly social in nature should be avoided as much as possible so that the valuable and useful technical type of conference or convention can be made available for attendance by utility personnel.

For groups of employees, as well as for individual employees, an effective and valuable method of pursuing specific education or training is by way of correspondence courses. Some correspondence courses can provide for and utilize excellent group participation in an educational effort, whereas almost all correspondence courses can be successfully pursued by individual employees either with or without help from others. The group correspondence course education method will require one or more strong educational leaders who need not necessarily be instructors or teachers, but should either have a firm foundation in the subject matter or be experienced in education and training techniques. The correspondence course method of education and training can be used for a wide range of subjects, from management and administration through supervisory skills, technical training, and high school education. The enrollment in correspondence courses by individual employees and groups of employees should be encouraged and supported by the water or wastewater utility manager as a part of the total employee improvement program of the utility.

The maximum benefit from an education and training program of a water or wastewater utility can best be achieved if the utility organization includes an education and training officer, either full time or part time, who will have the responsibility to administer the total program. Such an individual should serve in an administrative capacity within the utility organization, possibly in the personnel office if not in a separate education and training office. He must be able and willing to spend a considerable portion of his total work week to the development of education and training programs, the counseling of individual employees, and the administration and scheduling of the various types of courses which can be made available to and pursued by the employees of the utility. It is important, however, for the utility manager to provide personal time and support to the education and training program even if he has an officer to administer the program.

Employee Motivation

The purpose of a water or wastewater utility is to provide a high level of safe and reliable service to its citizens at minimum cost. To provide the required level of service it is the responsibility of the utility manager, as with managers of industrial and commercial organizations, to obtain maximum work performance and production from the employees of the utility. This work performance and production is in the form of both employee work effort and employee work results. Since water and wastewater utilities are not yet automated to the point where personnel can be considered only a minor factor in production, it is essential that the employees of a utility provide maximum work performance and production on a day-to-day basis. This production is in the form of treatment plant operations, laboratory analysis, clerical activities, accounting and purchasing, engineering, and all of the other individual efforts required to make the utility function effectively at minimum cost.

For many years in the early days of the industrial age, workers were given a specific job to do and were supervised by an individual who was expected to force as much work out of the workers as he could. During the course of these many years, it was found that force was not actually an effective method of obtaining high level performance from employees, and motivational management of employees came into being. The emphasis was moved from requiring the supervisor to force as much work as possible out of individual employees over to inward force or motivation on the part of individual employees to accomplish maximum quantities of work with the highest quality of work possible.

The motivational method of obtaining maximum performance from employees was developed on the basis of psychological studies of employees to determine their various needs, how those needs could be satisfied by management, and how the satisfying of those needs could lead to improved production by the employees. This chapter will not contain an in-depth study of employee motivation, but water and wastewater utility managers should have a reasonably good understanding of the hierarchy of needs and how to motivate various types and classifications of employees by satisfying the needs of those particular employees. By concentrating on each of the full spectrum of needs, from the food, shelter, and comfort needs of an unskilled laborer to the recognition and self-actualization needs of top-level management personnel, the utility manager and his administrative staff can be successful in achieving employee performance at a high level if the motivational management principles are adhered to on a continuing, day-by-day basis.

It is appropriate and completely understandable that the validity of employee motivation should be questioned. For many years, it was common practice for management to attempt to develop maximum employee productivity by force or some other type of coercive action. Even in some present-day utility organizations the principle of using various types of coercion to develop employee productivity is followed. With few exceptions, however, modern management, including the management of water and wastewater utilities, has proven that the determination of employee needs and the satisfaction of those employee needs have resulted in inward motivation of employees which has, in turn, resulted in increases in employee productivity. Even though the results of increased productivity in some of the water or wastewater utility activities may be difficult to measure, the overwhelming evidence is that employee motivation has resulted in productivity far beyond what has been accomplished by employee coercion.

Those who are skeptical of the validity and value of employee motivation are quick to point to civil rights legislation, civil rights activities, collective bargaining, and the increase in labor unions within water and wastewater utilities as a deterrent to management realizing benefits from employee motivation. It is unquestionable that the present day arena for motivation is such that it is often difficult for some employees to be motivated inasmuch as civil rights laws and activities, as well as labor union legislation and activities seemingly provide the satisfaction of most of the needs which employees previously tried to satisfy for and by themselves through increased production which resulted in monetary rewards or rewards of other types. With some of these lower-level motivational factors such as income and job protection seemingly provided by civil rights legislation and union contracts, some managers may wonder whether an emphasis on motivation will be effective. Within water and wastewater utility organizations the large majority of employees still have most of the higher levels of motivational needs which can be satisfied by effective utility managers. These needs include recognition for length of service, for safety record, for suggestions, for work performance improvement, and others. In addition, the possibilities of promotion and the many types of awards or rewards still available to utility managers for employees can be effective if used wisely by the utility manager.

Inasmuch as the largest expenditure within the budget administered by a water or wastewater utility manager is for wages and fringe benefits, it is essential that the manager concentrate as much time as is necessary on developing and maintaining a high level of employee motivation and a resulting high level of employee work performance. This is particularly important for professional and higher-level administrative and supervisory personnel, but it is also important for the many skilled, semi-skilled, and unskilled workers of the utility.

Communications Between Management and Employees

An examination of conflict between individuals and between groups of individuals would reveal that most such conflict has been the result of faulty communications. This is as true for the working relationship between management and personnel of a water or wastewater utility as for any other group of people. It is thus of extreme importance for the utility manager to maintain a continuing and effective method of communicating with employees for the mutual benefit of the utility and the employees.

An essential part of the communications from the utility manager to his employees is in the form of formal directives and guidelines. These formal directives set forth the goals and objectives of the water or wastewater utility together with specific requirements of the utility management and employees toward the accomplishment of those goals and objectives. The directives should include work rules and specific criteria for performance, provisions in the personnel manual which will impact the expectations of and benefits for employees, specific requirements of employees concerning a multitude of work-related or benefit-related actions, and a host of other formal requirements which management feels it is required to place upon employees of the utility or responsibilities which management assumes toward the employees. An effective set of management directives will form the basis of mutual trust and understanding between management and employees, and should give the employees a feeling of security, knowing specifically what is expected of them in the form of work performance and conduct.

An effective method available for employees of a utility to communicate with the utility management is an employee suggestion system. Through an employee suggestion system it is possible for employees to be formally involved in the development of revised work methods; improved facilities, safety, and employee benefit systems; and a wide range of activities which can be mutually beneficial for the employees and the utility. The suggestion system should be formalized to the point of continually receiving suggestions, preferably on a prescribed form and in accordance with established guidelines, reviewing the received suggestions on a regular basis by a formal suggestion review committee, responding to each employee for each suggestion received from him, and following through with some type of reward to employees who provide useful and valuable suggestions to the utility manager. Recognition of some kind, even if only a letter of appreciation, should be given to each employee who submits a valid suggestion, regardless of its value. Such recognition completes this type of communication as a two-way communication.

At those times when changes are envisioned in the utility facilities, processes, organization, work rules, the number of classifications of employees, or fringe benefits, or at times when employee problems are recognized, it is advisable for the utility manager to meet with groups of employees to explain the contemplated changes or the causes and probable implications of a specific problem which has been identified. By meeting with groups of employees, it is possible for the utility manager or his appropriate department heads to explain in detail and answer questions about changes which otherwise could be misunderstood, could result in conflict between management and workers, or could result in personal problems for employees.

Adequate advance notice of such group meetings with employees should be given to the appropriate employees, with information about the specific purpose of each meeting. The meetings should be structured as follows: management personnel first explain the problem or change as understood by the utility manager; then employees should be given the opportunity to ask whatever questions they may have concerning the subject or subjects of the meeting. All suggestions provided by employees at the meetings obviously should be given thoughtful consideration in the ultimate decision-making process concerning the subject of the meetings. The employee group meeting is another effective method of providing both communication from the utility manager to the employee and the opportunity for the employees to communicate directly with the utility manager or his representative about specific key issues.

Many private companies and public agencies publish a newsletter for the purpose of communicating important management information to the employees as well as to provide to employees the opportunity to learn more about the activities of fellow employees, the utility or company organization, and contemplated programs and problems. The newsletter gives many an employee one of the few opportunities he may have to see his name in print and receive publicity concerning some noteworthy achievement in a specific work assignment, utility sports team, or educational endeavor.

Inasmuch as the first line supervisor is the primary contact between management and the workers of the utility, it is essential that the management personnel of a water or wastewater utility provide for the development of maximum communication between supervisors and the workers they supervise. Regular meetings of the utility manager, his key management personnel, and his supervisors will enable the manager to promote improved communication skills of supervisors in dealing with their workers; will enable the manager to provide detailed explanation of changes in directives, working conditions, construction, and program plans; and will give the manager the opportunity to provide any other specific information which otherwise could not be adequately communicated to the workers. Improvement of the supervisory skills of supervisors can be accomplished by supplementing formal education and training activities for supervisors with regularly scheduled meetings at which actual supervisory problems as well as hypothetical supervisory problems can be discussed in detail away from the pressures of actual working conditions.

Another effective avenue of communications between utility management and utility employees can be provided by scheduling regular meetings between management personnel and union stewards or other personnel who represent nonmanagement personnel. These meetings can be used for airing problems or complaints presented by employees to union stewards, providing advance information of possible or probable changes in working conditions, changes in employment and promotion conditions, and any other pertinent information concerning the personnel, activities, or organization of the water or wastewater utility. The use of this type of meeting can avoid many grievances, can stabilize uneasy relationships between management and unions, can improve worker productivity, and can generally provide many communication benefits to the utility manager. This type of meeting will, at the same time, provide the individual employees with a feeling of closer contact with and more personal involvement in the management of the utility.

One of the most common means of communicating with employees is a system of regular announcements or posted bulletins which furnish meaningful information on the various activities of the utility. Employees are motivated to increased production when they feel that they are a vital part of the utility organization, have knowledge of the goals and objectives of the utility, and are included in the early dissemination of information concerning changes in programs or initiation of projects of the utility. Well informed employees generally can be expected to be more productive than employees who have relatively little knowledge of how their particular job performance is related to the successes, failures or progress of the utility.

The manager of a water or wastewater utility may not wish to use some of these specific types of communication with his employees, but it is essential that he be constantly aware of the need of employees to receive communications from management and that he provide that communication by some means. The opportunity for employees to communicate with management either directly or through supervisors, union stewards or other representatives must be provided by the water or wastewater utility manager as an essential part of his responsibility in providing management and direction for the total work force through the supervisory and management personnel of the utility.

Collective Bargaining and
Negotiation of Union Contracts

During the 1960s and 1970s the representation of employees of local governmental agencies by labor unions increased across the country and it is reasonable to assume that in the foreseeable future most skilled and unskilled employees of water and wastewater utilities will be represented by some type of labor union. It was to be expected that as employees of public agencies saw their counterparts in industry and private business receiving some degree of benefits from labor union affiliation they would seek some of the same anticipated benefits by forming or joining unions. Since increasing numbers of water and wastewater utilities will have at least some of their employees represented by unions, it is imperative that utility managers develop a knowledge of and experience in the many aspects of collective bargaining unit representation, union contract negotiations, handling of grievances in accordance with prescribed procedures, the arbitration of grievances which cannot be settled within the management structure of the utility, and the preparation for and continuation of operation of utility functions during strikes.

Regardless of whether the employees of a water or wastewater utility are represented by a labor union, it is reasonable to assume that they will resort to some degree of collective bargaining and that the utility manager will be required to negotiate with representatives of the collective bargaining unit. This will require that the utility manager understand the collective bargaining procedures and the requirements of management in contract negotiations with employee collective bargaining units. In most cases, it is desirable for the utility manager to have available for such contract negotiations the services of an individual who has extensive experience in collective bargaining, negotiations with collective bargaining units, contract negotiations, and all other aspects of collective bargaining. The cost to the utility for the services of such an individual is minor compared to the overall benefits which can be realized from the use of such an expert in the collective bargaining process.

It is important for the utility manager to realize that in union negotiations the basic intent of the collective bargaining unit is to obtain from management as much as possible in the way of wages, fringe benefits, and other benefits. A basic requirement of management is to understand the basis of the collective bargaining unit position and to be able to negotiate each item proposed by the bargaining unit on the basis of reason, cost, the prevailing level of wages and benefits in the utility field and in the community, and such other facts as may be required to negotiate objectively across the bargaining table with the union or other collective bargaining unit representative.

During the course of negotiating a labor contract with a collective bargaining unit it is imperative that the utility manager preserve the historical rights of management, virtually at any cost. Any relinquishing of the rights of management will be a disservice to the citizens served by the utility, and ultimately will require that management and the citizens served by the utility pay dearly to regain the management rights previously negotiated away. Some of the rights of management which without fail must be retained by management include the following: the right to manage the utility facilities; the right to schedule working hours; the right to establish or change work schedules or standards and direct the working force, including the right to determine the size of the working force; the right to hire, promote, and transfer personnel from one job to another; the right to suspend, discipline, or discharge any employee for just cause; the right to determine the location of business, including the establishment of new plants or divisions and departments and the relocation of any plant or division; the determination of necessary equipment maintenance; the allocation and assignment of work of employees; the control and use of utility property, materials, machinery, and equipment; the determination of safety, health, and property protection measures for utility personnel and facilities; the establishment, modification, and enforcement of reasonable rules and regulations which are not in conflict with the express provisions of the contract; the determination of the employees who are to be laid off or transferred because of a change in economic conditions or facilities operations; the transfer of work from one job to another; the establishment of new job classifications or the modification of existing classifications; introduction of new, improved, or different equipment; the placing of services, general maintenance, or construction with outside contractors or subcontractors; and the determination of the amount of supervision necessary.

Any citation in a labor contract of examples of those rights to be retained by management should be prefaced by a statement to the effect that, without limiting the generality of the statement in the contract concerning the rights of management, the term "Rights of Management" as used in the contract includes the examples set forth. Following the listing of examples of the rights of management, it is wise to include yet another statement to the same effect; for example: "The foregoing examples merely demonstrate by way of illustration examples of management rights which are retained by the utility, and it is specifically agreed that the enumeration of the management rights in the contract shall not be deemed to exclude other management rights not specifically enumerated."

The negotiation of a labor contract with utility employees and their union representatives must be based not only on the wishes and demands of the union representatives and employees, but also on the wishes and demands of management. The contract normally will include specific provisions pertaining to working conditions and hours, pay and fringe benefits for employees, the expectations of management concerning employee performance, a detailed procedure for handling grievances of employees against supervisory and management personnel or conditions, a procedure for arbitration of grievances which cannot be settled within the management procedure and many other provisions. Table 11-1 lists the articles of a sample contract containing many of the specific provisions which will often be incorporated into a water or wastewater utility union contract.

Table 11-1 / Sample Union Contract Provisions.

1. Union Recognition	12. Voting Leave
2. Union Security and Dues Payment	13. Holidays
3. Negotiation Committee	14. On-the-Job Injury Leave
4. Rights of Management	15. Seniority
5. Grievance Procedure and Arbitration	16. New Job Classifications
6. Hours of Work and Overtime	17. Working Conditions
7. Court Leave	18. Group Insurance
8. Emergency Leave	19. Safety
9. Military Training Leave	20. Pension
10. Vacation Leave	21. No Strike Provisions
11. Sick Leave	22. Term of Agreement
	Appendix. Rates of Pay

One of the most important provisions to be included in a union contract is a detailed procedure for handling grievances of employees, including the arbitration of such grievances. The grievance procedure should require that the grievant first must formally present his grievance in writing to his supervisor, who will be required within a specific number of days to respond to the grievance (first step). If the answer by the supervisor is not satisfactory to the grievant, the grievant may then formally submit the grievance to the personnel officer or similar official within the management structure of the water or wastewater utility for consideration at a hearing (second step). If, after this hearing, the grievance answer by the personnel officer or similar official is not satisfactory to the grievant, the procedure then normally provides for the grievant to take the grievance to a final hearing (third step) before the manager or other top management official of the utility. This final hearing will include evidence from both management and union representatives. Within a prescribed period of time after completion of the third step, the manager will make his final decision on the grievance and advise the grievant, the union representatives, and the management officials included in the grievance of his decision. If the grievant is not satisfied with the final decision by the manager, he has a specific period of time within which he may request that the grievance be submitted to arbitration in accordance with the arbitration provisions of the union contract. An important aspect of grievance arbitration, as well as other arbitration matters, is the method of determining who will arbitrate the matter. In one common arrangement, the contract will specify that a management arbitrator meet with a union arbitrator, and that together they select a third arbitrator who would then arbitrate the grievance. Another popular method of selecting the arbitrator involves the mutual agreement on five arbitrators to be considered for arbitrations. When an arbitrator is needed, the management and union representatives take turns deleting one of the names of the arbitrators listed on the list of five arbitrators until only one arbitrator remains. The remaining arbitrator is then charged with arbitrating the grievance.

Administration of Discipline

No organization, including any water or wastewater utility, can function properly and provide adequate service to its customers unless an effective system of discipline is employed within the organization. It is not normally necessary for employees to be dealt with severely as a result of rules infractions, but it is essential that all employees realize that if they violate work rules or directives of the utility they can expect swift, fair, and firm disciplinary action.

As a result of some employee grievances being upheld against supervisors and management personnel and also as a result of legislation such as the Civil Rights Act and various enactments of labor legislation as well as recent court rulings pertaining to Equal Employment Opportunity legislation and regulations, some employees have tried to take advantage of the various laws and regulations which were intended to protect their rights, but which are being interpreted by these employees as giving them special privileges. In many instances, employees who have been subjected to disciplinary action and have as a result, filed grievances only to have the grievances denied under the grievance procedure, have carried the grievance through the Equal Employment Opportunity and Civil Rights procedures seeking special consideration because of race, color, or sex. It is thus essential that utility managers have a complete understanding of the many opportunities employees have for airing grievances, and the applicable laws and regulations. The utility manager also must ensure that disciplinary action against employees is appropriate, is based on the known facts of the violation, and is well documented.

An effective procedure to be used in the administration of discipline which will provide fairness to both the individual employees and supervisors, and which will also stand the test of civil rights hearings and court actions is the procedure generally referred to as progressive discipline. The use of progressive discipline will provide that increasingly serious infractions of rules and repetitive infractions of rules will result in increasingly more severe discipline, and could ultimately result in termination of employment of the individual employee. Extremely serious cases in which an employee is guilty of criminal action, extraordinary negligence, insubordination or serious safety infractions should result in long-term suspension from duty or termination of the employee. In each such serious case, however, the utility manager must ensure that the severity of the infraction warrants the severe penalty. In other cases in which minor infractions of rules by an employee occur from time to time, the discipline should progress from a verbal warning to a written warning to be included in the employee's file, followed by a short period of suspension from duty without pay. Further infractions should lead to long-term suspensions or other more severe disciplinary action. If several of these steps of progressive discipline do not correct the attitude, conduct, and work performance of the employee, demotion or termination of employment would then be the final step in the progressive discipline.

The use of progressive discipline within a utility organization must be well understood by all levels of management and the supervisors who will be primarily responsible for the administration of discipline. It should be obvious to the utility manager that the proper administration of progressive discipline can be accomplished only if the supervisors are fully aware of the policy and philosophy of the utility manager concerning the administration of discipline and the use of progressive discipline in all of the routine as well as particularly serious disciplinary cases, and are indoctrinated and experienced in the important area of employee counseling. Unless the supervisors are well informed in all aspects of discipline administration, it is probable that disciplinary action which results in grievances will probably be overturned in arbitration, in civil rights hearings, or in court. Employee morale and productivity within the water or wastewater utility can suffer considerably unless the supervisors are capable of administering discipline properly, firmly, fairly, and in such a way as to ensure that disciplinary action will not be reversed by reviewing authorities. Full documentation of each specific disciplinary action is essential to effective discipline which can withstand arbitration, civil rights review, and court test.

As already mentioned one facet of the administration of discipline which needs improvement in most organizations is the documentation of disciplinary action and the reasons for the disciplinary action. Each infraction of rules by an employee should be documented by the supervisor of the employee if such offense warrants any type of disciplinary action including a verbal warning. In the event that progressive discipline does not correct the performance, attitude and conduct of a troublesome employee, it will be essential that each offense, each counseling effort of the supervisor, and each disciplinary action be well documented for review during the grievance procedure, during arbitration of a disciplinary action grievance, or at a civil rights or court hearing.

If the overall administration of discipline within a water or wastewater utility is to be successful, it is essential that clear and specific work rules be promulgated governing the work performance and conduct of employees of the utility. Even though the utility manager may feel that he can reasonably expect all of his employees to conduct themselves and perform their duties in accordance with good conscience and good taste, present-day laws and regulations together with social pressures require that the conduct of employees must be judged in accordance with specific work rules of the employer. It thus is essential that the utility manager promulgate, for his utility employees, work rules which specify the types of conduct which will not be acceptable to the utility management. Conduct which should be prohibited under formal work rules includes: insubordination; acts of disrespect toward management; making of false statements; falsification of records; reporting for work under the influence of alcohol or drugs; consumption of alcohol or drugs while on duty; deliberate misuse, abuse, or destruction of utility property; threatening, intimidating, coercing, or abusing other employees of the utility; unauthorized removal or theft of property or materials of the utility, of other employees, or of any other person; leaving the job without relief or authorized permission; failure to follow posted or published safety rules; sleeping during working hours; unauthorized absenteeism or lateness; and negligence or carelessness which results in danger, damage, injury to another person, or loss of utility property. Employees should understand that violation of the published work rules will result in disciplinary action.

A final point to be emphasized in connection with the administration of discipline is the absolute necessity of administering discipline uniformly throughout the organization. It is completely unacceptable for office employees and operators, or maintenance employees and laboratory workers to be disciplined on different standards. The utility manager must ensure that no department or division directive ever tends to establish disciplinary procedures which may conflict with the disciplinary procedures set forth by the manager or which may be different than that required for administration by other departments or divisions.

Discrimination

It is appropriate to preface any discussion of discrimination in employment with the general statement that any such discrimination on the basis of race, color, creed, sex, or ethnic background is bad and should not be permitted. This is particularly true in the selection, promotion, and discipline of employees of a public water or wastewater utility. In recent decades, great emphasis has been placed on certain types of discrimination, especially pertaining to employment of individuals. Many legislative and regulatory actions have accomplished a lot to reduce certain types of unacceptable discrimination, but in so doing have in some ways caused other serious discrimination.

The civil rights movement in the United States has existed for many years and has resulted in considerable change in the relationship of most employers with the several types of minorities. The manager of a water or wastewater utility is governed by civil rights legislation and regulations in all aspects of employee relations, including the selection and promotion of employees, disciplinary action, and the indoctrination of administrative and supervisory personnel in the complex requirements for avoiding unlawful discrimination against any employee for any reason.

In addition to the legislative requirements concerning discrimination against minorities, civil rights legislation and regulations require utility managers to ensure that females receive full consideration and equal consideration with males for job opportunities, promotions, and types of employment. It has become necessary for the utility manager to hire females for many jobs for which only males have historically been employed, and to provide additional facilities such as showers, locker rooms, and toilet facilities for segregating male and female employees where previously only male employees were of concern.

In some jobs which previously were assigned only to males, or virtually only to males, it has been found that female employees can perform the required work fully as well as male employees. In other jobs, it has been found that only occasionally can a female employee perform the same physical work expected of the specific job classification as is done routinely by male employees. In all of these cases, it is necessary for the utility manager to ensure that all prospective employees, regardless of race, color, creed, sex, or ethnic background will be given equal opportunity to be hired, to be promoted, to be transferred to a different job classification and to receive comparable pay for work performed.

No discussion of discrimination in employment can be complete without inserting precautionary comments concerning what can be referred to as reverse discrimination. The United States Congress and appropriate federal agencies, along with comparable state agencies, have provided legislation and regulations which at times virtually mandate the hiring of a member of a minority group or a female by certain employers, including water or wastewater utilities. Such requirements also extend to promotions, job assignments, and disciplinary action by management. These requirements may come in the form of quotas or percentages or in some other form, but however they are set they are reverse discrimination. The utility manager should be familiar with the prevailing equal opportunity requirements, know how to comply with the law and still base decisions concerning employees on fact, merit, and fair judgment.

It is imperative that managers of water and wastewater utilities require that their administrative and supervisory personnel be fully aware that they are not to resort to discrimination for any reason, and that discrimination by them will result in disciplinary action against them. The criteria for selection, promotion, or disciplinary action pertaining to any employee must be based totally on performance, conduct, and fulfilling the needs of the water or wastewater utility. Any other criteria can easily be identified as discrimination and will certainly cause serious problems for the manager.

Summary

The determining of staffing needs for a water or wastewater utility is a complex procedure and should be accomplished with a combination of thorough study of the needs of the utility, the advice of officials of other utilities, the use of staffing information compiled by the United States Environmental Protection Agency and other appropriate agencies, and consideration of the labor market and other local conditions.

The qualifications for individual job positions within the utility organization should be based on the actual skills required by the individual job classification within the utility. Advice and guidance can be obtained from other similar utility organizations.

Education and training programs are essential to successful personnel advancement within a water or wastewater utility, and all employees of the utility should be provided the opportunity to avail themselves of the benefits of the education and training program of the utility.

The understanding and application of the principles of employee motivation by personnel in all levels of management and supervision will result in maximum employee performance and utility success.

Effective methods of maintaining continuing communication between utility management and utility employees will result in improved employee morale, improved employee work performance, a minimizing of conflict between management and employees and a maximizing of the level of service to the utility customers.

With an anticipated increase in the unionization of employees of water and wastewater utilities, it is essential that the managers of such utilities acquire and have available either on staff or on a consulting basis reasonable expertise in the negotiation of collective bargaining agreements and all aspects of management relationships with collective bargaining units.

Managers of water and wastewater utilities can avoid many serious personnel problems with effective administration of discipline within their organizations. Successful administration of discipline requires that management and supervisory personnel understand the need to apply discipline uniformly throughout the organization, to employ progressive discipline, and to provide complete documentation of all disciplinary actions.

Discrimination in all personnel activities must be avoided within water and wastewater utility organizations. All hiring, promotion, transfer, discharge, and discipline of employees must be on the basis of facts concerning ability, skill, attitude, performance, and conduct of employees.

12

Selection and Use
of Consultants

It will often be necessary for a water or wastewater utility manager to secure and retain the services of consultants who can provide necessary expertise which is not, at the time, available within the utility organization. Both because of the sporadic nature of the actual need for expert personnel and because of the economic benefit of paying only for the limited amount of time personnel of special expertise may be needed, it is common practice for utility managers to employ consultants of several different types and levels of expertise.

It is a common occurrence for a utility manager to wrestle with the question of whether to hire an additional expert employee or to retain the services of a consultant. Each individual case must be decided on the facts of the situation, including the funds available, the expertise needed by and presently available to the utility manager, and the period of time within which the relevant work must be completed. In most water and wastewater utility organizations it is desirable for the manager to have on payroll a minimum number of certain professionals, such as engineers, to perform routine work, and then to retain the services of consulting engineers for special design work, studies, supervision of construction, and other engineering duties which are beyond the capacity or capability of the engineering staff of the utility. This is usually also true for other areas of professional expertise in which it would not be economical for the utility manager to pay an annual salary for a professional individual whose services may be needed only part-time throughout the year, for only several months of the year, or only for a single special project.

Only in rare situations will it be possible for the manager of a water or wastewater utility to fulfill all his management responsibilities without hiring consultants for specific purposes. It thus is imperative that the utility manager develop an effective capability for selecting and utilizing consultants.

Consultants Normally Required

The types of consulting services most often required by the manager of a water or wastewater utility are professional engineering and architectural services. Many studies and design projects are required for the planning and design of water and wastewater facilities, and most utility organizations do not have sufficient staff personnel with an adequate range of engineering specialization to provide a complete study or design of facilities. The same is true with respect to the need for architectural services, since most of the architectural work required by a water or wastewater utility involves structures which are a part of facility design and construction, and the architectural work would normally be included in the design work for which a consulting engineer would be responsible. Because of the lack of need for routine architectural work, few water or wastewater utility organizations include architects on their professional staffs.

Major facilities construction for water or wastewater utilities is usually financed with either revenue or general obligation bond issues. Adequate preparation for a bond issue requires the guidance of an individual or firm with extensive experience in the sale of bonds if the bonds are to be issued at a reasonable interest rate, hence it has normally proved necessary for the utility manager to retain the services of a professional fiscal consultant to handle bond issues as well as other matters. The services of a bond counsel with an outstanding national reputation will also be required for the preparation of the necessary bond resolution and other routine bond counsel services which are intended to make the issue attractive to bond purchasers at the lowest rate the various circumstances permit.

Although some water or wastewater utilities may be fortunate enough to have full-time legal counsel on staff, most can secure the required legal advice and assistance only from a law firm or individual attorney on a consulting basis. The most common and preferred method of using the services of a legal consultant is to pay the attorney to attend regular and special meetings of the governing body and provide additional legal services as required by the utility manager on an hourly fee schedule. In some cases, the utility will pay a specified annual retainer to have the services of the legal counsel on a demand basis, and then pay an additional hourly fee. Regardless of the manner in which legal counsel is provided for the water or wastewater utility, it is imperative that the utility governing body and manager have available to them, on an uninterrupted basis, the advice and guidance of the legal counsel. As discussed in Chapter 9, many liability situations confront utilities, and legal guidance is not only desirable but essential in most actions of the manager and the governing body.

The use of a public relations or public information consultant by a utility manager can include a wide range of services from complete reliance on the consultant for all public information or public relations activities to a limited use or even no use of such consultants except for major bond issue programs or some other special public information or public relations activities. Some utility organizations will rely on the manager, or one or more staff persons designated by the manager, together with the chairman and other members of the governing body, to perform the public information and public relations activities for the utility. Other organizations will place enough emphasis on public information activities to retain on staff a full-time public information or public relations expert who will not only prepare and issue news releases and perform all other routine public relations functions, but also develop and manage major bond issue campaigns and other major public information and public relations projects. Still other utilities will retain a firm which specializes in public information and public relations work to handle all or virtually all of these activities for the utility organization.

For those agencies which purchase and use computer equipment for process control or data processing purposes, it is usually necessary for the utility manager to retain, on a consulting basis, the services of a firm to provide computer or computer systems maintenance. Such a firm would be used to provide maintenance services for various computer software or hardware depending upon the expertise required, the expertise available within the utility organization, and the expertise available on a reliable consulting basis. As staff expertise is developed for computer software and hardware maintenance, less reliance can then be placed on consulting services.

Several other types of consultants may be retained by a utility manager to provide specific services. These include insurance consultants, who advise the utility manager on various types of liability and casualty insurance needed by the utility; auditors, who provide annual audits of the fiscal affairs of the utility as normally required by law, by contract, or by bond resolution; safety consultants, who provide a variety of services pertaining to safety; and actuarial consultants, who provide actuarial services in connection with the retirement plan, if any, which is sponsored by the utility.

These are the principal types of consultants which would normally be required by the managers of water and wastewater utilities. Other consultants will often be needed to provide other services from time to time.

Selection of Consultants

A good basic rule to be followed by managers of water and wastewater utilities in the selection of a professional consultant is to base selection completely on the consultant's ability to provide the quality and quantity of services needed, and not on price, unless the utility manager knows that all the consultants considered for a specific engagement are capable of providing and can be expected to provide services of equal quality. An example of equal quality of service is found in the selection of an auditor, where audit results are considered to be similar for all firms under consideration. Similarly, in the selection of an actuarial consultant it may be presumed that any actuary being considered would perform the same actuarial services for retirement programs or similar activities. For engineering, architectural, fiscal, legal, public information, and other similar professional services, the best procedure is to analyze the capabilities of each firm to provide the specific services needed by the utility manager and to select the firm which the utility manager considers best qualified for the specific service. After selection of the consultant on the basis of qualifications, the manager can negotiate the agreement and the fee for the required professional services.

In the selection of engineering consultants for performing wastewater facilities study or design work to be partially financed with funds from the federal government, the wastewater utility manager must comply with a requirement that selection of the consulting engineer must be based on a procedure prescribed by the U.S. Environmental Protection Agency. This prescribed procedure includes notification by trade journal or newspaper to consulting engineering firms about the specific project for which engineering services are needed; an evaluation of the credentials of firms which indicate interest in the project; and the selection of a firm based on the firm's probable capability of meeting the requirements of needed expertise, availability of personnel for the specific project, assurance of sufficient numbers of appropriate personnel for the size of the project, and recommendations from former clients. After selection of the consulting firm, the utility manager then is in a position to negotiate the engineering agreement and the appropriate fee with the selected consulting firm for the specific project.

The selection of engineering consultants should be based, as previously mentioned, on expertise, personnel availability, specific experience and similar factors which indicate the ability of the consultants to perform the required work satisfactorily. It is important that not only the proven expertise of the firm, but also the expertise of individuals within the firm be evaluated with respect to the specific project for which the engineering services are required. It is also important for the utility manager to ensure early in the consulting firm evaluation process whether an individual with the required expertise will be made available to the utility during the entire life of the project. The consulting engineering firm should be expected to commit to the project specific personnel, especially key personnel, who will provide continuity of required expertise and also continuity of communication between the utility and the consulting firm.

The selection of other consultants for professional services should be based on the reputation of the consultants as well as the quality and availability of qualified personnel within the firm to perform the required services, whether this be legal counsel services, insurance advice and guidance, or public relations and public information consulting services.

Because the several kinds of engineering services are the primary kinds of professional services required by a utility manager, it is important for the manager to ensure that his governing body understands clearly that the cost of engineering services is a small portion of the total cost of a construction project. Since, virtually all studies and designs result ultimately in facilities construction, each engineering fee should be considered in light of and as a part of the total cost of construction of which the studies or design are a part. In most cases of new facilities construction or extensive additions to facilities, the total engineering cost will range from about 5 to 10 percent of the total cost of construction. This total engineering cost normally includes preliminary planning, design, and professional guidance, usually including construction inspection, during the period of construction. Any effort to economize on the cost of engineering services by competitive bidding or in some other way will certainly result either in a less adequate construction project than would otherwise be achieved, or in all probability, in a total cost of construction considerably greater than if price were no factor in the selection of the consulting engineer.

If the water or wastewater utility manager selects professional consultants on the basis of the capability and anticipated performance of consultants, he can be at least reasonably certain that he will receive good professional performance by the consultants and still not pay an exorbitant price. It should be obvious that the scope of services to be performed by a consultant must be well defined in the professional services contract with the consultant. The negotiated fee should be adequate to ensure a reasonable profit for the consultant and still be low enough to not place an unreasonable financial burden on the utility customers. After the fee has been negotiated it is the responsibility of the professional consultant to perform the required work in accordance with the scope of services defined in the contract. Inasmuch as the reputation for good work performance is one of the most important factors used in the selection of a professional consultant, it should be obvious that it is of extreme importance for the professional consultant to perform well on each project. Keeping this fact in mind should enable the utility manager, without any particular difficulty, to select a good professional consultant, negotiate a reasonable fee with him, and then expect a good professional job from him.

Role of the Consultant

In considering what the role of the consultant should be it is appropriate to examine in some detail each type of consultant who may be retained by the water or wastewater utility manager. It is essential that the consultant and the utility manager understand and agree completely on the services which are to be provided by the consultant, the constraints within which the consultant is to perform his services, the services for which the consultant will be paid the basic fee, and identification of other services which may be provided by the consultant in accordance with a specific request to do so by the utility manager and for which the fee will be specifically agreed to by the manager and the consultant prior to performance.

The role of the consulting engineer will be found to vary greatly according to the specific needs of a water or wastewater utility. The engineer may be retained for a technical study of the physical condition of facilities, the capacity of facilities, any operational problems, or the adequacy of rates and charges; he may be retained for the preliminary planning or design of facilities; or he may be retained to inspect or supervise the construction of facilities. The utility manager has many occasions to call upon a consulting engineer. When a consulting engineer has been retained to perform an engineering study of any kind it should be his responsibility to perform the required study within the limitations and to the extent specified in the professional services contract. Engineering study contracts normally require that the engineer, upon completion of the study, submit to the utility manager a report which sets forth the findings and conclusions of the study together with recommendations for action by the utility.

Whenever a consulting engineer is retained for the design of a water or wastewater facility the engineering contract will usually require the consulting engineer to provide a preliminary design. After receiving approval of the preliminary design and being instructed to proceed with the final design, the engineer will be required to perform the final design work. The final design work normally will include the preparation of final plans and specifications ready for advertising for construction bids. In most cases, the consulting engineer who performed the design work will also be responsible for overseeing the project through the steps of advertising for construction bids and the awarding of the construction contract. For purposes of economy and project continuity it generally is advisable for the manager of a water or wastewater utility to retain the consulting engineer who designed the facility for the administration of the construction contract and inspection of the construction.

In the administration of any construction contract the consulting engineer is the representative of the owner, and as such must represent the owner in all matters pertaining to the construction of the facility. It is the additional responsibility of the consulting engineer to mediate conflicts between the owner and the contractor, and in so doing, to provide a degree of professional protection to the contractor. Inasmuch as the consulting engineer is the agent of the owner, however, his primary responsibility obviously must be to the owner.

Legal and fiscal consultants are normally needed by water and waste-water utilities as has already been explained. The legal consultant can serve on a retainer basis for continuing routine legal work or can be retained only for specific legal actions such as condemnation of land, assistance in the preparation of bond resolutions, legal defense, or advice in matters pertaining to water law and water rights, or to defend the utility in specific legal actions brought by or against the utility. Fiscal agents likewise can be retained solely to handle the fiscal matters pertaining to particular bond issues or other special fiscal activities, or can be retained for routine fiscal advice as required by the manager or governing body of the utility. In each instance the legal or fiscal consultant and the utility manager must agree upon and completely understand the specific role of each with respect to their duties, responsibilities, and authority.

Public information and public relations consultants can be retained for the purpose of preparing routine news releases on a regular basis and to provide guidance to the utility manager on relatively routine matters. These consultants also can be retained solely for specific public information or public relations activities associated with specific bond issue campaigns, major land acquisition, facilities construction projects or at such other specific times as the governing body of the utility feel that the public image of the utility requires improvement or special public notice.

Other consultants such as the auditor, safety or insurance consultant, or consultants of any other type similarly may be retained on a continuing basis for routine guidance or on a job-by-job basis to perform specific services in connection with specific activities of the utility. In each case the role of the consultant must be specifically described in a contract and understood and agreed to by the consultant and the utility manager.

Whenever a water or wastewater utility manager thinks that he may have a need for a consultant for some particular activity, he should first determine whether the need is real. This determination will result from an analysis of the work to be accomplished, the quality and quantity of expertise available on staff, and the type of consultant needed if the necessary expertise is not available internally. The manager should then establish the general scope of services required of the consultant and determine specifically the proposed role of the consultant and the required relationship of the consultant to the governing body, the manager, the general public, and other agencies. The consultant should then be selected on the basis of maximum anticipated benefit to the utility in terms of performance, reliability, and cost. After the consultant has been selected, the manager should negotiate the specific scope of services and the terms under which the consultant will be retained by the utility. It cannot be emphasized too strongly that whenever feasible the consultant should be selected on the basis of his qualifications before the fee for the work is negotiated.

Consultant Contracts and Fees

Each consultant who is retained to perform specific services for a water or wastewater utility should be expected to perform those services in accordance with a signed contract between the governing body of the utility and the consultant. Legal counsel who may serve on a continuing basis without specific duties may perform his services in accordance with an official resolution of the governing body and under the general guidance of the governing body, the chairman of the governing body, or the utility manager. Each consultant contract must set forth clearly the services to be furnished by the consultant, the persons to whom the consultant will be responsible, and the basis and method of payment to the consultant for services performed.

The fee to be paid to a consultant for providing specified services to a water or wastewater utility can be based on any of several methods of payment: lump sum; cost plus a fixed profit; actual costs times a multiplier; or, in the case of a design contract, percentage of estimated or actual cost of construction. The determination of the method to be used in setting the payment to a consultant should be based on the type and extent of work to be performed, the detail with which the total scope of services can be identified, applicable laws and regulations which may prohibit or discourage certain methods of payment, and the type or method of fee payment customarily used by and within the profession represented by the specific consultant.

Payment to a consultant in the form of a lump sum fee is generally the easiest and least controversial method if the scope of services can be easily established and agreed to. When the consultant completes his assignment he receives final payment in the agreed amount and the transaction is completed. For those consultant services contracts which provide for lump sum payment, it is usually necessary for the utility manager to negotiate a change in the consultant fee each time the scope of services is materially changed for whatever reason. However, at any stage the governing body and manager of the utility know what the total cost of the consultant's services will be, while the consultant knows specifically what services he will be required to perform. Where the consultant has considerable difficulty in estimating in advance his own costs, he must set his fee high enough to ensure that if his estimates are low he will not perform the specific services for the client at a net loss or inadequate profit to him. This is an important disadvantage of lump sum contracts; in essence, the burden or risk of an underestimate by the consultant is paid for by the water or wastewater utility. No consultant is in business as a public service but rather to produce income for himself, so fee negotiations necessarily will result in a fee which will ensure the consultant that he will have a reasonable expectation of making a net profit on all or practically all of his jobs.

The cost-plus-fixed-fee method of payment normally provides for the consultant to be reimbursed for all of his payroll costs, including fringe benefits and overhead, together with all other appropriate expenses, plus a fee which is set at a specific dollar amount and which will represent the profit to be earned by the consultant. A disadvantage of this method of payment obviously is the fact that the utility manager usually does not have a good estimate in advance of the total cost to the utility of the consultant work under the contract. On the other hand, this disadvantage can be lessened by requiring the consultant to reveal to the utility manager his estimate of personnel time and associated costs together with his other estimated costs and then, from that information, setting a maximum total fee which is not to be surpassed without separate negotiation between the utility manager and the consultant. An advantage of the cost-plus-fixed-fee method of payment is that it is not necessary to have the scope of services as specifically and clearly defined as for the lump sum method, and any required changes in the scope of services during the term of the consultant contract will not require negotiation of a new fee unless the maximum fee limit will probably be reached.

A method of consultant fee payment similar to the cost-plus-fixed-fee method is the cost-times-multiplier method. With this method of payment, the consultant and client will agree that the total fee will be the total cost to the consultant times a factor such as 2.0, 2.2, or some similar agreeable factor. For this method of payment the fee is calculated on the basis of the total payroll cost including all fringe benefits, the other consultant expenditures incurred specifically for the project, and the agreed upon overhead costs. The total cost to the consultant thus computed would be multiplied by the agreed upon factor to arrive at the total fee. In the methods of payment which involve cost to the consultant plus a fee or times a multiplier, it is essential that the consultant furnish to the utility manager his itemized costs for examination before final payment of the consultant fee.

The most common method of payment to a consultant, specifically a consulting engineer, included in a contract for the design of facilities has been by percentage of construction cost. Design fee curves such as those developed by the American Society of Civil Engineers have served for many years as the basis for water and wastewater utility managers' negotiations with consulting engineers for the appropriate percentage of construction cost which would be paid as a fee to the consultant for his design, preparation of plans and specifications, assistance in the receiving of bids, and awarding of the construction contract as well as certain specified engineering duties during construction. In the great majority of the construction projects for which this percentage method of payment has been used it has proved to be a satisfactory, fair, and reasonable method of payment. During the 1970s, as a result of certain federal regulations and a concern on' the part of some officials that the consultant may design more costly facilities than necessary for the purpose of increasing his total fee under the design contract, an increasing proportion of design contracts for construction of wastewater facilities included provisions for payment to the consulting engineer of a fee which would be either a lump sum fee for specified duties or some other method such as cost times a multiplier.

During the conduct of work under most consulting engineer contracts it is necessary for the scope of services to be changed for any of many reasons. When the scope of services is found to require change, the utility manager must negotiate with the consultant the specific changes in the scope of services along with a change in the fee schedule. As with most other contracts which are amended by negotiation, the utility manager can expect the additional fee for a change in the scope of services of a consulting engineer contract to be relatively high as compared to the basic fee, hence changes should be avoided as much as possible.

It should be obvious that the preparation of an adequate, well-defined scope of services by the utility manager is an important part of contract administration responsibilities. This requires that the utility manager and his key staff personnel devote considerable time and effort to negotiating with the consultant the specific work required. Any time and effort which the utility manager and his staff may save at the time of preparation and negotiation of the scope of services to be provided by the consultant will probably be expended many times over at a later date when disagreement may develop either over work which the client feels the consultant did not accomplish but was required to do, or over payment for work which may not have been well defined originally.

Summary

Water and wastewater utility managers can expect to require the services of many consultants including engineers and architects, attorneys, fiscal agents, auditors, public information and public relations specialists, computer specialists, and safety and insurance consultants.

The selection of consultants to provide professional services should be based, to the maximum extent possible, on capability rather than on cost or the amount of the fee to be paid to the consultant.

The role of the consultant should be well defined, should be set forth in detail in the contract between the utility and the consultant, and should be based on ultimate agreement on such roles as determined by negotiations between the utility manager and the consultant.

Fees to be paid for consulting services should be based on whichever method of determination of fee best fits the type of work, applicable regulations, complexity of the work, difficulty of determining the specific scope of services to be performed, and other appropriate factors.

Contracts for consulting services should be prepared in such a manner as to minimize the need to amend the scope of services or any other part of the contract with the consultant after the contract has been executed.

13

Contract Administration

The manager of a water or wastewater utility is responsible for and involved in numerous activities which can be categorized as contract administration. Included in the scope of contract administration are contracts for the purchase of equipment, materials, chemicals, and tools; construction contracts; contracts for professional services including engineering, architectural services, fiscal services, legal services, and others; and contracts for miscellaneous services of many types. Each of these types of contract will be discussed in some detail in this chapter.

Purchase Contracts and Procedures

Most public agencies, including water and wastewater utilities, are required by law or by other mandate to purchase their materials, chemicals, equipment, and tools on the basis of competitive bidding by vendors and manufacturers. Even without a legal requirement to purchase on the basis of competitive bids, such a competitive bidding procedure is normally advantageous to a water or wastewater utility for the purpose of providing sufficient competition between vendors or manufacturers to result in the lowest feasible purchase prices for the water or wastewater utility. Depending upon the dollar amount of a given purchase, the purchase procedure may require sealed bids; written proposals or bids without any formal bid opening or evaluation procedure; telephone price quotations from several vendors; or single requests for prices from individual vendors.

Many of the purchases by a water or wastewater utility involve single items or groups of items which are required at a given time to replenish the stock for operations or maintenance purposes or for specific projects, whereas other purchases are required on a continuing basis throughout the year. Examples of the latter type of purchase would include such items as gasoline, diesel fuel, chemicals, and office supplies. The purchasing procedure for the latter type of purchases can be simplified and the work load minimized by taking competitive bids annually on those items which would be required on a continuing basis throughout the year rather than soliciting bids each time additional purchases of those items are needed. The purchase of items on an annual bid basis is recommended for as many purchase items as feasible in the interest of economy of cost and time.

Most public agencies, including water and wastewater utilities, normally are required to seek competitive bids for purchases and services, but even so there is often a certain amount of latitude available to the water or wastewater utility manager for purchases within specific ranges of costs. Whereas formal sealed bids will usually be required for the purchase of items of considerable cost, say, $3,000 or more, the purchasing agent may be authorized to accept written bids on an informal basis for the purchase of items of a cost of less than $3,000. He may be authorized to request verbal bids for much lower-cost items, say, items which cost less than $200, and he may be authorized to bypass bidding procedures completely if a given purchase item is available from only a single source or if emergency conditions exist.

In those cases in which purchase items reportedly can be obtained from only one source, it is essential that the utility manager ensure that this is not because the specifications are unreasonably exclusive or restrictive. He should also ensure that the selected source is truly the only source for purchase of the required item and that the purchase of the item without following any type of competitive bid procedure is actually legal for the specific water or wastewater utility.

The use of verbal bids, customarily obtained by telephone, is often an acceptable bid procedure for the purchase of items whose value or cost does not exceed some small dollar amount, say, $200 or $300. For the soliciting of verbal bids, it is usually sufficient for the purchasing agent to call three or four vendors for their prices on specific purchase items. It is essential, however, that the prices be documented along with the name of the company, the person who gave the price and the date on which the price was received.

For the purchase of items which have a higher cost, up to, say, $2,000 or $3,000, informal written bids should be requested. As when formal sealed bids are used, it is imperative when the utility manager or his purchasing agent uses informal written bids that the bids comply with the specifications for purchase as well as all of the bidding and purchasing requirements of the utility.

When requesting and receiving formal sealed bids, usually for all purchases which will cost more than $2,000 or $3,000, it is necessary for a prescribed formal advertising and bidding procedure to be followed by the manager and purchasing personnel of the water or wastewater utility as well as the bidders. Such a procedure must include such features as a notification of the prescribed date and time of bid opening, the form of bids or proposals, the purchase specifications, quality testing requirements, the furnishing of bid security, and other prescribed requirements.

It should be obvious that even though competitive bids are desirable for most purchasing purposes, especially for purchasing items which involve the expenditure of large amounts of money, it is not reasonable to require a formal bid procedure, with the accompanying expenditures of personnel time and money, to be followed for small purchase items which may have values of only $25 or $50. It also is normally not in the best interest of the citizens served by a water or wastewater utility to restrict the utility management in their purchasing procedures to the extent that personnel time is wasted or that less than the best quality of purchased items will be received or more than the best price paid. In the interest of best serving the utility customer, and for the protection of the water or wastewater utility manager, his purchasing agent and other individuals involved in the purchasing procedure, it is essential that a purchasing manual be formally approved and adopted by the governing body of the utility. In such a purchasing manual should be set forth the procedures to be followed for each type of purchase according to purchase prices, single or multiple sources of supply, and as many other special conditions as can be contemplated. Among the provisions in a purchasing manual should be: requisition procedures, bidding procedures, ordering procedures, and procedures for the sale of surplus property and scrap material.

Requisition procedures normally should require the following:

1. A purchase requisition must be submitted for each purchase order issued
2. Each purchase requisition must include a statement by a department head or person of similar authority that the purchase is necessary to the operation of the utility and is properly charged to the stated budget activity
3. Each purchase requisition must bear the approval by the utility manager or governing body of the utility depending on the spending authority limit of the manager (this approval may be delegated by the manager for purchases which cost less than certain amounts set by the manager)
4. Special provision should be made for those purchases or services which are authorized by formal adoption of the budget or by other special action of the governing body of the utility (such as utilities services, professional services, insurance, and maintenance)
5. Special provision should be made for purchases of small cost ($25, $50, or similar limit) by petty cash or by use of a special short-form requisition

Purchase requisitions should be submitted to the purchasing agent or other appropriate person in the purchasing activity on standardized forms which provide for:

1. Requisition number
2. Date of requisition

3. Name of budget activity to be charged
4. Explanation of the need for the items being requisitioned
5. Suggested vendors
6. Place of delivery
7. Date of desired delivery
8. Quantity and unit of each item
9. Complete description of each item
10. Estimated unit and total costs of each item
11. Account number to be charged
12. Statement signed by the requisitioning department head that the requested items are necessary for the operation of the utility and are properly chargeable to the indicated budget activity
13. Statement is signed by appropriate fiscal officer of the utility to the effect that an unencumbered balance is available in the indicated budget activity to cover the cost of the requested items
14. Certification by the utility manager that the required approval of the manager or the governing body has been granted (for purchases which cost more than some established amount such as $200 or $500)

For the requisitioning of items from the central stores of a utility, a stores requisition should be required in a manner similar to that for the purchasing of items. Such a stores requisition is simpler than a purchase requisition, but it should include such information and signatures as the following:

1. Date issued
2. Stores requisition number
3. Account number charged
4. Maintenance request number, if applicable
5. Items issued (quantity, description, unit cost, total cost, inventory credited)
6. Signature of employee issuing items
7. Signature of employee receiving items

The purchasing procedure of a water or wastewater utility should include formally adopted procedures for soliciting competitive bids for the purchase needs of the utility. Such bidding procedures may include provisions similar to the following:

1. A notice of specific purchasing needs of the utility shall be sent to suppliers and contractors in the manner which can best achieve maximum competition among bidders and resulting maximum economy to the utility
2. *Purchases up to $500.00.* Bidding is not required for purchases up to $500.00 in price. All purchases, however, shall be transacted in such a manner as to achieve maximum economy to the utility
3. *Purchases from $500.00 to $2,500.00.* Whenever possible, at least three telephone quotations will be obtained and a record kept of such quota-

tions for purchases from $500.00 to $2,500.00. Decisions as to whether a written quotation is necessary will be left to the buyer's discretion
4. *Purchases from $2,500.00 to $5,000.00.* Written quotations are required to document purchases estimated to cost from $2,500.00 to $5,000.00. Written quotations will, whenever possible, be solicited from at least three potential bidders. Care will be taken to ensure that the purchase is made in compliance with the terms and conditions prescribed by the utility
5. *For purchases exceeding $5,000.00.* Formal written proposals must be requested for purchases of supplies and services estimated to exceed $5,000.00. By public advertising or other appropriate means, notice of the invitation for bids shall be adequately circulated, and whenever possible, proposals shall be requested from at least three responsible bidders
6. Formal or informal procedures may be waived by the manager of the utility when supplies or services being purchased are within one or more of the following categories:
   a. *Limited availability,* namely, supplies or services which are obtainable for practical purposes from only one single source
   b. *Urgently required,* namely, supplies or services which must be purchased to contend with emergency situations
   c. *Determined by practicality,* namely, supplies required in respect to preferences based on particular individual usage or professional advice

Telephone quotations, when appropriate and authorized, should be requested and documented on a telephone quotation form which would include the following:

1. Vendor name
2. Name of vendor representative
3. Date of each quotation
4. Quantity, description, unit price, and total price of each item needed
5. Recommended award
6. Remarks
7. Signature of the employee who obtained the quotations and recommends the award.

When informal written bids are solicited they should be submitted on appropriate forms which include:

1. Bid number
2. Date of bid invitation
3. Date of closing of bids
4. Instructions for submission of bids
5. Item number, quantity, unit, description, unit price, and total price of the items needed
6. Statement that all prices should be quoted f.o.b. the utility address
7. General conditions (on reverse side of form)
8. Statement that the supplier agrees to furnish the listed items at the indicated price

9. Date and signature
10. Company name of the supplier

Formal written bids should be received on appropriate forms which include, in addition to the information included on the form for informal written bids, a notice of the time and place for opening the bids, special conditions for purchase or contract work, and information concerning the basis of acceptance of bids by the utility and a notice of award to the successful bidder.

The general conditions which govern the submission of formal and informal written bids should include the following requirements and assurances:

1. Bids must be submitted on the forms provided and received prior to the time of bid opening
2. Bids must be sealed and the envelopes in which the bids are placed must be clearly marked as bids for specific purchases, services, or construction
3. Bids must be signed by a duly authorized official of the bidding company
4. Bids may only be withdrawn on written or telegraphic request prior to the time fixed for the opening of bids. All bids will otherwise be considered to be firm and open to acceptance or rejection for a period of 45 days or some other reasonable time from the date of opening
5. The utility reserves the right to reject any and all proposals, to waive any informalities or irregularities in the proposals received, and to accept the proposal from the lowest responsible bidder affording the greatest economy to the utility
6. Property which, upon delivery, or work which, upon completion, does not meet the specifications and standards prescribed herein, or property which has been damaged in transit, may be rejected by the utility at the bidder's risk and expense
7. When bids are solicited for a certain article or an article equal thereto and a bidder intends to furnish an article which he considers equal to the one named, he must specify in his bid the trade name and the grade of such article, together with such engineering data and technical literature as may be required by the utility
8. When the words "no substitute" are used in the description of an article, quotations will be accepted only on the article described
9. The utility reserves the right to request from any bidder free samples for testing to determine the quality of proposed materials or products
10. All disputes concerning grades and quality of merchandise or work shall be determined and settled by the manager of the utility, or other persons properly authorized by the utility manager
11. The utility is exempt from all federal taxes under the provisions of Chapter 32 of the Internal Revenue Code (Registration Number A-123456) and from all state and municipal sales and use taxes under the provisions of the appropriate state provisions

12. On the bid form which is provided, the unit price of each item must be given in the column headed "Unit Price." This price must be submitted for the particular unit of measurement specified in the column headed "Unit." In case of error in extension of prices, the unit price will govern. All prices must be extended and totaled

13. When a date or dates are set for the delivery of property, delivery must be made on or before the date or dates called for in the specifications or call for bids, or the utility will have the right to cancel the contract for the purchase of the bid items and to purchase equivalent bid items at market prices for immediate delivery and hold the bidder liable for any increase in the price over and above the prices offered in the proposal

14. The bidder will protect, defend, and save harmless the utility against any demand for payment for the use of any patented material, process, device or article that may enter into the manufacture or construction or form any part of the property covered by the offer; and the bidder will further indemnify and save harmless the utility, its officers, agents, and employees from any claims, suits, or actions of any nature growing or arising out of the acceptance of the offer or the exercise of any rights under the contract for sale created thereby

15. Payment for all items shall be made within 30 days or some other reasonable time after delivery to and acceptance by the utility

16. In connection with work under the contract, the contractor agrees not to refuse to hire; or to discharge, promote, or demote; or to discriminate in matters of compensation against any person otherwise qualified, solely because of race, creed, color, national origin, or ancestry; and further agrees to insert the foregoing provision in all subcontracts under the present contract

17. Except where otherwise expressly stated in the contract documents, all terms therein employed shall have the same definition as set forth in the appropriate state regulations

18. The supplier or contractor agrees to abide by all the laws, regulations, and administrative rulings of the United States, the State, and the utility, securing all necessary licenses and permits related to the purchase covered by the bid proposals

19. Trade discounts (i.e., deductions from the list price of goods) shall be computed by the bidder into the unit price and the extended price for each line item before these prices are entered in the appropriate columns of the proposal form. The utility will not assume the responsibility for making these computations

20. Cash discounts (i.e., deductions from total invoiced amounts for payment of bills within specified time limits) quoted for periods of less than thirty (30) calendar days or other reasonable time from the date the invoice is received by the utility will not be considered in the evaluation of any proposal

21. No contract will be awarded to any person, firm or corporation that is in arrears to the utility, upon debt or contract, or that is a defaulter, as

surety or otherwise, upon any obligation to the utility, or that may be deemed irresponsible or unreliable by the manager of the utility

22. All participating bidders, by their signature thereunder, shall agree to comply with all of the conditions, requirements, and instructions of the bid as stated or implied therein. Should the utility omit anything from the bid which is necessary to a clear understanding of the bid, or should it appear that various instructions are in conflict, then the bidder shall secure written instructions from the manager of the utility at least forty-eight (48) hours prior to the time and date of the bid opening

23. *Noncollusive bidding certification.* By the submission of the bid, the bidder certifies that:

   a. The bid has been arrived at by the bidder independently and has been submitted without collusion with any other bidder; and

   b. The contents of the bid have not been communicated by the bidder, nor, to its best knowledge and belief, by any of its employees or agents, to any person not an employee or agent of the bidder or its surety on any bond furnished herewith, and will not be communicated to any such person prior to the official opening of the bid.

For those purchases for which special conditions are required, the bidding documents should include such special conditions, which will provide, among other requirements, the following information:

1. Detailed specifications
2. Estimated or fixed quantities
3. Term of contract
4. Shipping instructions
5. Freight allowance
6. Manufacturing facilities information
7. The right of the utility to renew or extend the contract
8. Requirement for the successful bidder to maintain local inventories
9. Trade-in information
10. Requirements for servicing of vehicles or equipment prior to delivery
11. A warranty from the bidder that the purchased items are free from defects in materials and workmanship

Special conditions which would normally be required for construction contract bids will be covered in this chapter in the discussion of construction contracts.

In the preparation of detailed specifications for purchasing contracts, several important considerations must be given particular attention. These considerations include the need to determine which is more important to the water or wastewater utility, lower price or specified desired features; the selection solely on the basis of low purchase price only if the various manufacturers or suppliers can be expected to provide products of comparable quality; the preparation of specifications in such a way as to favor a particular desired product; and whether the specifications should be prepared in such a way as to not only meet the minimum needs of the utility as to product quality, but to also reflect the confidence of the utility manager in either the supplier or the manufacturer as to performance in complying with purchase contract provisions, or to give major consideration to the level of service provided by the supplier. It is often difficult in preparing specifications to decide whether several desired features in a product will be more important to the utility than the consideration of low price. The preparation of specifications for purchasing obviously must include advice from the operations and maintenance personnel who will be using or operating the product or will be required to keep the purchased equipment maintained, repaired, and in reliable operation. In those cases in which all of the manufacturers or suppliers of a specified product can be expected to provide a product in accordance with the specifications, the price of the product is then the only governing feature unless quality and reliability of service by local representatives is also of concern. At times it has been common practice for a purchasing agent to prepare purchasing specifications so as to enable the preferred supplier of a specified product to always submit the low bid price. Although this procedure is not illegal per se, it should be practiced with extreme caution to avoid undermining the entire competitive bidding process. It is often necessary to analyze many factors other than merely the price and the required minimum quality of the product to be purchased. As stated previously, any desired qualities of a product as well as the required minimum qualities must be given proper consideration, as must the reliability of suppliers in fulfilling their contract obligations, the degree to which the capability of suppliers to meet their obligations is dependent on the manufacturer of their product, and the quality or reliability of service to be provided by the local supplier or the service agency for the supplier of the purchased product.

The evaluation of bids for purchase contracts can range from a simple comparison of bid prices to an in-depth investigation not only of the prices, but also of the degree to which the proposed product would comply with the specifications as to quality and delivery requirements, as well as the probable quality and reliability of service from the supplier for such purchase items as vehicles or other equipment. If the purchase specifications have been properly prepared it should be relatively easy to identify those bids which propose to furnish acceptable products. For those bidders who have complied with the specifications it is then appropriate, after determining the three bidders with the lowest bid prices, to investigate each of those low bidders with respect to reliability of supply of the product and similar products, service reliability of the supplier if service is a concern, and any other appropriate factors such as local supplier versus out-of-state or out-of-region suppliers and delivery dates. This investigation of the several low bidders should rely both on the experience within the present utility and the experience of other water and wastewater utilities and similar agencies. Extensive testing of materials to be furnished under a purchase contract or minimum testing of such materials should be provided for in the specifications for the bids, and should be conducted in accordance with specifications before payment is made to the supplier.

To ensure that the purchasing procedures which are followed within a water or wastewater utility are uniform, fair, and legal, the utility manager should prepare a purchasing manual which clearly sets forth all procedures required for purchases by the utility, and the purchasing manual should be approved and adopted by the governing body of the utility. The purchasing manual should set forth the bidding requirements for various types and monetary values of purchases, the identification of individuals in the organization who may authorize purchases of various monetary values, and the other information needed to govern the entire purchasing activity of the utility. The several specific provisions which are recommended for inclusion in the purchasing manual are covered in considerable detail earlier in this chapter.

Construction Contracts

Any discussion of the subject of construction contract administration, if it is to be complete and meaningful, should include consideration of an entire construction project from the time when the utility manager, his planning, engineering, or operations personnel and the governing body of the utility understand the need for a construction project up to the time of completion of construction, final acceptance of the completed work by the utility, and final payment to the contractor. Such a discussion must include the study and design of the project; the bidding documents, bidding procedure, and awarding of a construction contract; administration of the contract during construction, including construction inspection, change orders, and progress payments; and the final acceptance of the construction and final payment to the contractor for the completed construction.

As discussed in Chapter 6, it is essential to the continuing successful providing of service to the public by water and wastewater utilities that the utility manager develop a long-range planning program which will enable the management personnel of a utility to identify problems and the need for facilities construction many years in advance of the time when the facilities construction will be needed, and to plan facilities to meet those needs. An adequate long-range planning program will enable the utility management to commence, at an early date, a systematic study of specific problems and the need for specific facilities construction. Delays of many kinds can be anticipated in connection with most construction projects, and it is thus essential that the long-range planning effort and the predesign study of facilities planning work be scheduled and accomplished years in advance of the time when construction is needed. Prior to the time that federal laws and regulations unreasonably constrained water and wastewater utility planning, a major construction project could be planned, designed, and built in three to four years. With present regulatory delays the same major construction project will take a minimum of seven years for the same planning, design, and construction.

Coordination of the long-range facilities plan and a corresponding financing plan must be accomplished and maintained on a continuing basis if the funding of construction projects is to be available at the time when construction contracts are to be awarded.

Effective coordination of the study and preliminary planning of a proposed project construction and the design of the project can be best accomplished by utilizing the services of a single consulting engineering firm for the two phases. For many possible reasons, however, it may be necessary or desirable for two separate consultants to be used for the study and design phases of a construction project, or the engineering staff of the water or wastewater utility may do either the study or all or part of the design of the facilities. Regardless of how the study and design phases are performed, it is necessary for the utility manager to ensure that the study and design phases are well coordinated.

Effective construction contract administration requires that the utility manager devote sufficient effort and resources to all parts of the construction project, including design of facilities; preparation of plans, specifications and other contract documents; advertising of formal sealed bids; receiving and public opening of the bids; the evaluation of the bids; awarding of the construction contract to the bidder who has submitted the lowest and best bid and which is considered to be in the best interest of the water or wastewater utility; inspection of the construction and the materials of construction; preparation and payment of progress payments; final inspection of the construction; and final payment to the contractor. If a federal or state grant-in-aid provides a portion of the construction cost, it is imperative that the utility manager ensure that all aspects of the contract administration comply with appropriate laws or regulations pertaining thereto. The consulting engineer who has been retained for design of the proposed facilities should be expected to prepare the plans and specifications for the construction contract bidding, and also should perform the other contract administration duties under an engineering agreement with the utility.

The contract documents to be prepared by the consulting engineer or the engineering department of the utility for the purpose of soliciting construction bids should include:

1. Advertisement for bids
2. Instructions to bidders
3. Proposal form
4. Form of contract
5. General conditions
6. Detailed specifications
7. Plans and drawings
8. Bond forms and instructions
9. Insurance instructions
10. Miscellaneous certificates information

The advertisement for bids should be placed in journals, magazines, newspapers, and other publications which are normally used by contractors who can be expected to bid on the specific construction work. The advertisement should include information stating the utility name and address to which bids are to be submitted; the deadline date and time for submission of bids; a brief description of the construction work for which bids are requested; the locations at which plans and specifications may be examined and obtained; the amount of deposit for obtaining plans and specifications; the bid security requirements in the form of a bid bond or certified check; the performance and payment bond requirements; the requirements for successful bidder compliance with federal and state requirements; notification of any prebid conference location, date, and time; the requirements concerning compliance with conditions of employment and minimum wage rates; and the absolutely essential statement that the utility reserves the right to reject any or all proposals, to waive informalities or bidding irregularities, and to accept that proposal which is in the best interest of the utility. In addition to placing the advertisement in various publications, it is usually wise for the water or wastewater utility manager to send copies of the advertisement to those contractors whose bids are particularly desired and should be specifically solicited.

The instructions to bidders are intended to be a guide to the prospective bidders in the preparation and submission of their bids as well as to advise prospective bidders of the method of determining the successful bidder and the procedure for awarding the contract. The instructions to bidders should include information about the format of the contract documents; special comments about the specifications language being "command" type sentences; reference to the general description of the project as contained in the advertisement for bids; a statement concerning interpretation of documents to be made only by the engineer in the form of addenda; clarifying statements, if needed, concerning the plans; information concerning the type of proposal (unit price, lump sum, or a combination); instructions for preparation of proposals; consideration of sales and use taxes in the bid prices; instructions concerning the submission of proposals; telegraphic or written modification of proposals; permissable withdrawal of proposals; bid security in the form of cash, certified check, cashier's check, or bid bond; return of bid security; the requirement for prospective bidders to become personally familiar with the job site and conditions rather than to rely on information furnished by the engineer as opinions rather than fact; method of contract award; basis of award of the contract; execution of the contract especially pertaining to the time within which the contract must be executed after notice of award; performance and payment bond requirements; the implications of failure by the bidder to execute the contract and furnish required bonds; and a statement to the effect that the work covered by the contract must be completed within the agreed time, with reference to delays and extension-of-time provisions of the general conditions. The instructions to bidders should include the statement that the utility reserves the right to reject any or all bids, to waive irregularities in bids, and to accept the bid which in the opinion of the utility governing body is in the best interest of the utility.

Although the proposal form may vary considerably among different water and wastewater utilities and among different consulting engineers, it should include certain minimum information, including:

1. Title, setting forth the construction project for which the proposal is being submitted
2. Name and address of the water or wastewater utility and manager
3. Project number
4. Bidder name
5. Date
6. A statement that the bidder declares that the only persons interested in the proposal are those named within the proposal, that the proposal is, in all respects, fair and without fraud, that it is made without collusion with any official of the utility, and that the proposal is made without any connection or collusion with any person making another proposal on the same contract
7. A statement that the bidder has carefully examined the contract documents for the construction of the project; that he has personally inspected the site; that he has satisfied himself as to the quantities involved, including materials and equipment, and conditions of the work involved, even though the description of quantities of work and materials is brief and intended to indicate only the general nature of the work; and that the proposal is made according to the provisions and under the terms of the contract documents
8. A statement that the bidder agrees that he has exercised his own judgment regarding interpretation of subsurface information and has utilized all data which he believes pertinent from the engineer, the owner, and other sources in arriving at his conclusions

9. A statement that the bidder agrees that if his proposal is accepted, he will, within 10 calendar days (or some other suitable period of time) after notification of award, execute the contract with the owner in the form of contract included in the contract documents; and will at the time of execution of the contract deliver to the owner the performance and payment bonds required by the contract documents; and will to the extent of his proposal furnish all machinery, tools, apparatus, and other means of construction and do the work and furnish all the materials necessary to complete the work in the manner, in the time, and according to the methods specified in the contract documents and required by the engineer thereunder

10. A statement that the bidder agrees to begin work within 10 calendar days after the date of written notification to proceed and to complete the construction, in all respects, within the specified number of calendar days set forth, after the date of notification to proceed, in writing, by the owner

11. A statement that in the event the bidder is awarded the contract and shall fail to complete the work within the time limit or extended time limit agreed upon, as more particularly set forth in the contract documents, liquidated damages shall be paid to the owner at the rate of $500 per day (or some other suitable amount) in default until the work shall have been furnished as provided by the contract documents, with the provision that Sundays and legal holidays will be excluded in determining days in default

12. A statement that one contractor will be selected for award of the contract as defined under "Basis of Award" in the contract documents

13. A statement that the bidder agrees that he has reviewed and understands the method of contract award as defined in the contract documents, and has prepared his bid accordingly

14. A statement that the bidder must bid all parts of the construction work or that bidders may bid portions of the construction work and in the former case, that proposals that do not include a bid for all parts of the work will be nonresponsive and may not be considered

15. A statement that the bidder agrees to accept as full payment for work proposed under the contract, as therein specified and as shown on the plans based upon the bidder's own estimate of quantities and costs, the indicated amounts. The proposal form should then include the appropriate descriptions and spaces for bid amounts in both numbers and words

16. A statement which identifies the surety which would provide the performance and payment bond if the bidder is awarded the contract

17. The name and address of the bidder

18. The names of the principal officers of the corporation or partnership submitting the proposal, or of all persons interested in the proposal as principals

19. Appropriate signatures, titles, and date

The form of contract should be brief and should include by reference all of the contract documents as part of the contract. Statements which should be included in the contract form are those which: set forth the date of the contract and the official names of the utility and the contractor; commit the contractor, in return for payment to him by the owner, to perform the required work under the contract in accordance with his proposal and in full compliance with the contract documents; include by reference all contract documents as parts of the contract; commit the owner to pay to the contractor the bid amount in accordance with the proposal and the other contract documents; commit the contractor to complete the work within the specified time and to accept as full payment the amounts determined by the contract documents and based on the proposal; commit the contractor to indemnify and save harmless the owner from defects in materials or workmanship for one year (or some other reasonable period of time); and commit the contractor to completing the work by the specified deadline date or pay the specified liquidated damages. Signatures of the authorized officials of the utility and the contractor together with their titles and date as well as a signature of an attorney who approves the form of contract should conclude the contract form.

The general conditions are intended to provide guidance to contractors concerning contract requirements of a general nature which are normally applied to any construction contract with which the water or wastewater utility may be involved. Such requirements may include the following:

1. *Definitions*, including terms such as contract documents, owner, contractor, bidder, engineer, surety, specifications, notice and work
2. *Contract documents*, including intent of the contract documents, specification format, discrepancies and omissions, alterations, documents to be kept on job site, copies to be furnished and ownership of drawings
3. *The engineer*, including authority of the engineer, duties and responsibilities of the engineer, rejected material, right to retain imperfect work, lines and grades, shop drawings, and many other similar descriptions of specific requirements and procedures
4. *The contractor and his employees*, including subcontracting, performance and payment bond, insurance, permits and licenses, superintendence, sanitation, safety precautions and others
5. *Progress of the work*, including beginning of the work, schedules and progress reports, prosecution of the work, owner's right to do work, delays and extension of time, liquidated damages, use of completed portions by the utility and cleaning up
6. *Payment*, including partial payment, change orders, claims, release of liens or claims and final payment

These are only some of the provisions which should be included in the general conditions. Review of the general conditions contained in the contract documents for a major utility construction project will provide guidance in the specific general conditions to be used in the contract documents for any construction project of a water or wastewater utility.

In addition to the general conditions section of the contract documents there often will be required a section of special provisions, which largely will be amendments to the general conditions for the specific construction project. The special provisions may apply to ownership of drawings, lines and grades, shop drawings, performance and payment bond, insurance, special requirements of a state law concerning public contracts, correction of defective work after final acceptance, contract completion date and final payment, to mention only a few.

The detailed specifications which are prepared for the control and guidance of contractors for a specific construction project will include the detailed requirements for site preparation and landscaping work, the materials and workmanship for concrete, masonry, metals, other materials, finishes, painting, equipment, doors and windows, mechanical and electrical work, and any other details of the construction work. In the case of both detailed specifications and general conditions, it is essential that the language of the specifications be clear and concise, with an absolute minimum of interpretation necessary. The primary concerns of the water or wastewater utility manager in reviewing the specifications should be the use of acceptable-quality materials, acceptable workmanship, minimizing conflict between plans and specifications, and maximizing the clarity of wording of the specifications with resultant minimizing of need for interpretation of the specifications by the engineer. Even though the utility manager must rely on the expertise and judgment of the engineer, he should require and provide a reasonable degree of review of the specifications by staff engineering personnel.

The construction plans normally will be thoroughly checked for accuracy and completeness at least once in the office of the engineer, but the utility manager should also provide at least a minimum level of checking of the plans by in-house engineering personnel. The manager should also ensure that the documents clearly state that in any cases of conflict between plans and specifications, the specifications will govern.

The remaining portions of the contract documents, namely the bond forms and instructions, insurance instructions, and miscellaneous certificates information, are normally of standard form, and can be conveniently adopted from other construction contracts. In such adopting, however, it is essential that the forms and information be reviewed and revised, if necessary, to correspond to current requirements and conditions specific to the contract at hand.

When the plans and specifications along with the other parts of the contract documents are completed, and the date has been selected for receiving and opening the construction bids, the utility manager must provide for the advertisement for bids to be published in local newspapers, local and regional contractor journals, and, depending on the size of the project and the probable interest in the project by other than local contractors, in other regional and national technical journals and magazines. Care must be taken to ensure that adequate time is provided to prospective bidders to properly investigate the project and prepare their bids. Even one or two weeks of extra time for bid preparation can make the difference between unreasonably high bids and acceptable low bids. The location for public opening of bids should be selected on the basis of adequate space for the number of bidders and other people expected to attend the bid opening, convenience of the bidders as well as the utility personnel, and any other factors which will have any impact on the orderly conduct of the formal bid opening. The utility manager should plan in advance to be represented at the bid opening by a key staff person, the engineer, and the legal counsel for the utility. As bids are received by the water or wastewater utility the outside of the envelope in which they are received should be plainly marked with the date and time received and with the initials of the person who received the bids. The bids should be delivered unopened to the chief engineer of the utility or some other person designated by the utility manager to hold such bids for safekeeping.

On the prescribed date of bid opening, the chief engineer of the utility or other designated presiding officer, together with the legal counsel of the utility, should be available to receive bids up to the designated deadline time in the room selected for the bid opening. Bid tabulation forms should be made available to all persons in attendance. The water or wastewater utility should be represented at the bid opening by the consulting engineer for the project as well as the presiding officer and the legal counsel for the utility. At the scheduled time of bid opening the presiding officer should state that the time for bid opening has arrived and that no additional bids will be accepted. He should also state the order in which bids will be opened. Usually this is done in random order as they happen to occur, or the bids can be opened in alphabetical order or any other order chosen. The bids should then be opened by the presiding officer one at a time, the bid bond or certified check acknowledged and stated to either be in order or not acceptable by the legal counsel of the utility, and then the bid prices, time required for construction and other required proposal information read aloud. At the conclusion of the reading of the bid information for the last of the proposals which have been received for the specific project, the presiding officer should indicate the apparent low bidder, and should state that each bid will be checked for accuracy, that references will be investigated, and that bidders will be informed of the results of the bid evaluation by a certain approximate date, if feasible.

The bid evaluation process would commence immediately after the bid opening, and should be completed as soon as feasible after the formal opening of bids. Even though the evaluation of bids should proceed without delay, the evaluation process should be accomplished with extreme care, so as to ensure awarding of the construction contract to the bidder who submits not necessarily the lowest bid, but the bid which is in the best total interest of the utility. All dollar figures included in each proposal must be checked for accuracy of addition, subtraction, and multiplication. In unit price proposals it is particularly important to ensure that the extensions for each item are correct as well as the total bid amounts. The bid evaluator must be certain that each bidder has not inserted special conditions in his proposal, in which case the bid should be rejected for being unresponsive to the bid advertisement. An important part of the bid evaluation is the contacting of clients for whom the contractor has performed construction work in the past with the intention of determining the quality of completed work, the past history of meeting construction deadlines, the extent to which the contractor filed claims for additional payment, the ease with which the resident engineer and others have worked with the contractor, and any other information which would enable the water or wastewater utility manager to determine whether the low bidder, or one of the low bidders, could reasonably be expected to perform a satisfactory construction job if awarded the contract for the present project. The financial status, key personnel, and appropriate construction equipment of the low bidder also should be investigated, as well as any other historical or current activities of the bidder. The evaluation of bids and the several lowest bidders should be thorough enough to provide sufficient information with which the utility manager can recommend awarding of the contract to the bidder who has submitted the lowest and best bid.

After the utility manager has submitted his recommendation for award of the construction contract to the utility governing body, and the contract has been awarded, the bidders should all be notified of that action. At that time the bid bonds or bid deposits should be returned to all but the three low bidders. The bid bonds or bid deposits of the three low bidders should be retained until a contract has been signed with one of the low bidders, but the bid bonds or deposits should then be returned immediately after the formal contract has been executed by the utility and the successful bidder. The contract should be executed as soon as possible after the required performance and payment bonds and the required certificates of insurance have been supplied to the utility manager. The notice to proceed with the construction work should then be sent immediately to the contractor who has executed the contract so that work under the contract can proceed, usually within 10 days after the notice to proceed has been sent to the contractor.

During the course of the construction under the contract the engineer and the utility manager should do everything feasible to expedite progress payments to the contractor. The same should be true of final inspection, final acceptance of the work, and final payment to the contractor. Progress payments to contractors are usually made monthly in accordance with progress payment estimates prepared by the engineer in consultation with the contractor and in accordance with the schedule for payments to be made by a specified date each month. A reasonable and adequate percentage of the total payment due the contractor at any time should be retained by the owner so that the contractor at that point of completion of the total construction will have more money withheld by the owner than the value of work which the contractor still is required to perform for the owner under the contract. It is important to the owner that the retainage be adequate to ensure proper progress of work to completion of the project, but the amount of the retainage also should be fair to the contractor. It is usually adequate for 10 percent of the earned payment to be retained by the owner up to the time that more than 50 percent of the construction project has been completed, at which time the retainage usually may be reduced to a smaller percentage, such as 5 percent of the earned payment. The retainage then should be held by the owner until such time as the construction project has been completed and accepted by the owner. In some construction activities it has been proposed by contractors that in lieu of retainage of payment by the owner, the comparable funds be placed in escrow for the purpose of ensuring the earning of interest for the contractor for that portion of the payment earned by the contractor, but which is still being withheld as a form of retainage. Inasmuch as the purposes of the retainage of a part of payments are to protect the owner against work not being completed in a timely manner, and to force the contractor to complete the work expeditiously, it is normally wise for the retainage to be withheld by the owner with no provision for escrowing the funds. This is particularly important for the case in which the contractor has been slow to complete portions of the job, has failed to meet deadlines, and in other ways has shown a tendency not to diligently pursue any part of the required work.

For relatively large construction projects which require a considerable amount of equipment or materials to be stored for perhaps many months before installation, it is common practice for the owner to make payment minus retainage to the contractor for equipment and materials delivered to the job site but not yet incorporated into the work. When this early payment is made to the contractor, it is important that the contractor be held responsible for the security of the equipment and materials, and also that equipment warranties not go into effect until such time as the equipment has been installed, placed into operation, and approved by the engineer.

From the time when the construction contract has been executed by the contractor and the water or wastewater utility up to the time of final payment after completion and final acceptance of construction of the project it is of extreme importance to the utility manager for change orders to the contract be held to an absolute minimum. Change orders can be required because of changed conditions such as unsuspected underground obstructions or changed conditions, because of an unexpected need to redesign a portion of the project after award of the contract, because of a decision to increase or decrease certain construction quantities, and for a host of other reasons. As a general rule, it can be assumed that a relatively small number of change orders during the course of a construction project is an indication of good engineering, and a large number of change orders an indication of engineering of a lesser quality. There obviously are many exceptions to this general rule because of differences among owners and the degree to which they change their minds, differences among contractors, and differences among the many physical conditions which may be encountered during the course of a construction contract. Since the negotiating of prices and time extensions for change orders always places the owner at a disadvantage, it is to the financial advantage of the owner to ensure that the change orders to a construction contract be minimized.

The quality of construction performed for a water or wastewater utility under a construction contract will be largely dependent on the quality of inspection provided by the water or wastewater utility engineers during the course of the construction. A well designed project with good inspection will yield good construction results, while the same design with inadequate inspection can often result in questionable or faulty construction. The inspection services may be provided by staff personnel of the utility, if available, or by personnel provided by the consulting engineer who performed the design work and who usually furnishes a resident engineer in charge of the construction project. In the event that the utility manager has qualified inspection personnel available for performing the inspection services they should be utilized as fully as possible on the appropriate contract construction projects.

Inspection of construction projects is never a prudent place for a water or wastewater utility manager to cut costs or economize. Any money which may be saved by reducing inspection coverage on a construction project will usually result in problems and extra costs in the total construction of the project or in future operations or maintenance of the completed facility. Inasmuch as construction inspection costs are a minor part of the total construction cost of a project, usually only 1 to 5 percent of the total cost, it is always wise to provide the quantity and quality of inspection necessary to provide adequate coverage for all phases of a construction project at all times while the contractor is working. As an important part of the construction project inspection activity, the utility manager must ensure that the relationships between the utility manager, his engineering personnel, the consulting engineer and his resident engineer, the inspectors, and the contractor are well defined and understood by all parties to the construction contract. The construction inspector and contractor should understand completely that the only purpose of the inspector is to enforce compliance with the plans and specifications for the construction of the project. If an interpretation of plans or specifications is necessary, it is only the resident engineer or his superiors who can provide such interpretation. At no time should a construction inspector be permitted to interpret plans or specifications.

In the performance of his inspection duties, a construction inspector should at all times be firm in his enforcement of the plans and specifications, but at the same time he should be fair to the contractor, his superintendent, and his workmen. Even though the loyalty and primary responsibility of the inspector are to the owner, the water or wastewater utility, it is sometimes necessary for the inspector, through the resident engineer, to intercede on behalf of the contractor when the plans, specifications or special conditions jeopardize the fiscal position or construction capability of the contractor. It should always be kept in mind that a water or wastewater utility which treats contractors unfairly or unreasonably can expect such treatment to be reflected in higher bid prices on future construction or a lack of acceptable bids on future construction.

In the area of quality control of materials of construction, it is essential that the consulting engineer, on behalf of the utility manager, provide a level of quality control which can ensure that all of the materials which are incorporated into a construction project meet the minimum standards set forth in the specifications for the construction project. By rigidly enforcing the provisions of the plans and specifications pertaining to the quality of materials, workmanship, and equipment, the inspector can ensure an acceptable total construction project for the water or wastewater utility.

Service Contracts

The water or wastewater utility manager will often find it necessary to consider contracting for maintenance, custodial, operational, and other types of services normally performed by employees of the utility. For such services to be performed by contract rather than by utility employees, it must be shown that the contractor could provide the service at equal or less cost or will provide a higher quality or greater reliability of service than would be possible with employees of the utility, or that the service actually could not be provided by employees of the utility. The types of service contracts which should normally at least be given consideration by a utility manager include among others custodial services, maintenance of equipment, maintenance of facilities, grounds and landscaping care, uniform rental and laundering, maintenance of meters, security, quality control, and laboratory services.

Professional services have already been discussed in Chapter 12, and will not be considered here.

Whenever the utility manager is thinking of contracting for services which could instead be provided by personnel of the utility, it is essential that he or his staff determine that the contracting of the service will result in a total net benefit to the water or wastewater utility. The evaluation of total net benefit should include the total cost to the utility, including wages and fringe benefits; the quality and reliability of the service, especially pertaining to possible strikes by employees; availability of equipment; skills not available among utility employees; and any other considerations which may be relevant.

It is best for the utility manager to award service contracts on the basis of competitive sealed bids received from prospective contractors. The competitive bids should be received in response to an advertisement for bids and specifications which set forth in detail the quality and quantity of services to be rendered by the contractor. Upon receipt of the bids, the utility manager should ensure that adequate evaluation of the bids is performed by appropriate utility personnel, with consideration being given to past experience of the bidder, financial responsibility of the bidder, personnel and equipment available for the proposed service, and all other aspects of the bidder's organization and his capability of performing the service in accordance with the specifications at the bid price. In addition, it is important that the utility manager determine the impact on the utility if, after a service has been contracted on the basis of cost, in a subsequent year it is determined that the service should be performed by utility employees rather than by contract. Consideration also must be given to possible conflicts with labor unions which represent utility employees, the hardship which might be borne by utility employees laid off as a result of the contract work, and security problems which may result from contracting certain work.

Summary

The administration of contracts for purchases, construction, and services will require a large amount of personal attention, concern, and time on the part of a water or wastewater utility manager.

The purchasing procedures of a water or wastewater utility should be thoughtfully developed with the primary intent of securing purchases of the best available products at the minimum cost for the utility, and these procedures should be formally adopted by the governing body of the utility. These formally adopted purchasing procedures should be described in a purchasing manual and made available to all personnel of the utility who are involved in purchasing. The purchasing of materials, equipment, and fuel should be accomplished on the basis of competitive bids whenever feasible.

Insofar as feasible, materials and equipment purchases should be in accordance with detailed specifications which are clear and easily understood by prospective bidders and which are fair to the prospective bidders. In the preparation of purchasing specifications the utility manager should solicit the advice and technical guidance of supervisors who will be responsible for using the purchased products.

Construction contracts must be awarded on the basis of competitive sealed bids, and such contract awards must be in accordance with clear, well defined advertisements, instructions to bidders, general conditions, plans, and specifications. The contract documents should be prepared in accordance with accepted good practice within the engineering profession. Administration of construction contracts should include high-quality construction inspection as well as firm but fair relationships between the utility manager, the consulting engineer, and the contractor. Progress payments should be paid to the contractor promptly, change orders should be held to a minimum, and the contractor should be required to adhere to the requirements of the contract documents throughout the course of the construction work.

The water or wastewater utility manager should contract for nonprofessional services when such contracted services can be shown to result in a net benefit to the utility. The competitive bidding process should be employed for this type of contract work the same as for purchasing or construction.

14

Safety Considerations

Accidents of virtually every kind have been experienced by employees of water and wastewater utilities, and many of those accidents have been extremely costly to the utilities in terms of money, lost time, and individual suffering. Because of the severe impacts of accidents on the operation and finances of their organizations, utility managers have an obligation to develop and implement a continuing program of safety consciousness and practice which can result in maximum safety for employees, maximum participation by employees of the utility in such a program, and a minimizing of the number and severity of accidents involving not only the employees of the utility, but the general public as well.

The Occupational Safety and Health Act passed by Congress in 1970 caused much controversy throughout the United States because of some of the requirements which appeared unreasonable, but passage of the Act did increase, throughout the nation, the awareness of occupational types of accidents and the need to better control such accidents. Similar legislation enacted by state legislatures has had similar effects on employers and employees alike in both the public and private sectors. It is imperative that maximum effort be exerted by management to improve and maintain safety in water and wastewater utilities, with the belief and understanding that no reasonable cost of intensifying the safety programs of such utilities is too much in comparison to the benefits to be received from such safety programs.

Construction Safety

Many situations and conditions exist on and around water and wastewater utility construction sites for many kinds of accidents. Regardless of whether the construction work is performed by contractors or by the employees of a water or wastewater utility, these accidents can be sustained by the workmen, utility employees who must work in close proximity to the construction, and also the general public. It is of extreme importance that the manager of a utility insist on the development of and adherence to appropriate safety precautions in all parts and phases of any construction activity, from the groundbreaking ceremonies through all of the various construction activities up to the point of completion and final acceptance by the utility of the construction.

Most water and wastewater treatment facilities construction is accomplished by contractor forces, and in this case the safety of the construction personnel is less of a concern than in the case of construction projects performed by personnel of the utility. Still, even though the personnel of the utility may not be involved to any large extent in construction, it is imperative that the utility manager insist on maximum safety precautions being taken throughout the construction site and throughout the construction contract period by the contractor and his representatives. Detailed safety requirements should be included in the construction contract documents, and rigid enforcement of such safety precautions must be provided by the utility manager as well as by the consulting engineer who is responsible for contractor compliance with the construction plans, the specifications, and the various provisions of the other contract documents. In addition to safety precautions about the construction site for safety of construction personnel, it is necessary for the contractor to provide for maximum safety of utility officials and inspection personnel about the site as well as the safety of the general public who may drive vehicles or walk near the site. The protection of adjacent property also must be the responsibility of the construction contractor, and all safety precautions which may be required must be understood by the contractor and must be rigidly enforced by the inspection personnel of the utility or the engineer.

Small construction projects such as minor sewer construction projects or minor alterations to pump stations or other facilities have many of the same accident opportunities as are present with major treatment facilities construction. It is important that the utility manager recognize the vulnerability of personnel and property to accidents within and near the sites of small construction projects as well as in the vicinity of major construction projects. Since the safety hazards are basically the same in both major and minor construction projects, the same safety precautions and enforcement of safety precautions are required for small construction projects as for major construction projects.

Each construction project which involves the installation of a water main or a sanitary sewer is beset with numerous safety hazards, and special emphasis must be placed on the adherence by construction personnel to safety precautions throughout the progress of such construction. The various types of heavy equipment used for trenching, materials handling, backfilling, and other work related to water and sewer main installation present continuing conditions which are conducive to accidents, and recognized safety precautions must be enforced at all times when such construction equipment is in operation. Special emphasis must be placed on requiring workmen near the construction work to wear hard hats; on providing barricades to keep workmen and other persons out of the area through which cranes, backhoes, and similar equipment may rotate or move during their operation; on keeping workmen out from underneath pipe or other materials being lowered into the trench; and on requiring warning devices on equipment which will sound when the equipment is moving in reverse.

The necessity for construction personnel to work at the bottom of trenches constitutes a continuing potential for serious injury or death in the event of major trench cave-ins and other similar accidents. The requirement for adequate shoring of trenches, the prohibition of heavy loads of any kind on the ground surface adjacent to trenches, and the providing of special protective shields for trench workmen should be specified clearly in the construction specifications, and such requirements must be rigidly enforced throughout the course of the construction work. In the event that ground-water is found to constitute a potential hazard to workers on the job, it will be necessary for appropriate pumping equipment and in many cases well-point equipment to be provided and used for adequate removal of the trench water and the proper disposition of such water.

The protection of vehicular traffic and pedestrians in the vicinity of any trenching work should be a high-priority responsibility of the contractor or construction superintendent and should be insisted upon by the utility manager. In addition to providing for the safe movement of traffic and pedestrians past and across water and sewer main installation projects, it is important that barricades, signs, flashing beacons, and other safety equipment be employed to prevent to the maximum extent any access by children and other people to the actual construction site.

The protection of existing public and private facilities and other utilities which are adjacent to water or wastewater facilities construction projects must be mandated in the construction specifications and these specific safety provisions must be conscientiously enforced. Many examples from the past can be cited to illustrate the danger of undermining buildings which are located adjacent to water main or sanitary sewer construction excavations, and of undermining water, sewer, gas, and electrical utility lines across or adjacent to excavations, as well as the resulting damage to or destruction of such existing facilities or utilities. The safety precautions around construction sites thus must include extensive protection measures for property and structures adjacent to construction, other utilities which may be located within the work area, and also any equipment or materials of construction which have not yet been incorporated into the work.

The safety of motorists and pedestrians in and around construction sites must also be a primary concern of the utility manager, the consulting engineer, and the contractor. The water or wastewater utility manager must ensure that the construction site safety precaution requirements will provide adequate protection for all persons who may be in the vicinity of the construction not only during the hours of construction work, but also during the time when construction work is not in progress. The importance of the protection of motorists who must travel adjacent to, through, or across a construction site makes it necessary for such motorists to be provided with advance warning a sufficient distance from the construction site to alert them to the construction danger, to the need to reduce speed, to lane closures or constrictions, to detours, to flagmen or to any other similar safety measure. Detours should be well identified throughout their length with sufficient signs, arrows, barricades, and flashers to guide each motorist through with a minimum of confusion or delay and as rapidly as safety will permit. All warning or detour signs and equipment must be of such type as to be clearly visible and understandable at night as well as in the daytime. Flagmen should be well trained and should fully understand the importance of their actions. If lane closures require motorists to transfer to other lanes, a warning must be given to motorists far in advance of the point at which the lane transfer must be made. A good general rule to be observed for motorist protection should be that no surprise should confront the motorist at or near the construction site.

No effective, continuing accident prevention program can be expected to grow spontaneously out of the reactions of utility managers and employees to hazardous situations and serious accidents. The program must be the result of extensive advance planning for safety. Experience with hazardous situations and resulting accidents obviously will be valuable in the development of an effective accident prevention program, but this experience must be only one part of the total safety planning effort. Each construction project warrants individual investigation and evaluation of its potential safety hazards, for only on this basis can the utility manager plan for the prevention of accidents to workmen and to motorists and citizens, who with inadequate warning might unknowingly approach or enter a construction site or an area in which potentially dangerous utility operations may be encountered.

A few examples of what would constitute inadequate or ineffective safety precautions may be more meaningful in this regard than examples of adequate or effective safety precautions. Examples of some inadequate or ineffective safety precautions include: a lack of barricades and battery-powered flashing signals around an excavation or, if they have been furnished, barricades or flashing signals which are not properly maintained; the lack of sufficient numbers of signs to warn motorists and pedestrians of excavations and construction activities; the lack of warning signs not only alerting motorists to danger when approaching construction areas which will require extra caution, but also providing a warning about mandatory turns or the need to change from one traffic lane to another; the lack of flagmen when flagmen are needed for slowing, stopping, or guiding traffic; the lack of safety equipment to be worn by workmen; inadequate shoring of trench excavations; inadequate pumping or wellpoint equipment for excavation dewatering; inadequate inspection of cables, hydraulic systems and other critical parts of construction equipment; and unsafe scaffolding or ladders. Many other examples could be added. The advance planning of safety precautions around a construction site should include a great deal of thoughtful deliberation by the manager and the engineers involved for the purpose of identifying those types of accidents which could occur around the particular site. This identification of possible accidents should then be translated into specific safety precaution requirements to be included in the construction specifications or general conditions.

Safety in Operations and Maintenance Activities

As with construction activities, it is necessary for the water or wastewater utility manager to conscientiously and continually plan safety into all of the operations and maintenance activities of his utility as well. As stated previously, the reaction to serious accidents can never be a satisfactory substitute for effective, coordinated advance planning of safety as an integral part of the total planning and management of operations and maintenance activities.

An area of employment activity in which safety precautions required on the job will prove to be equally beneficial in the personal lives of employees involves the operation of motor vehicles. The strict requirement for safe and courteous operation of the utility's vehicles and equipment by utility employees will make the employees increasingly aware of the need to drive safely at all times whether on the job or away from the job. The manager of a water or wastewater utility cannot afford to assume that those employees who possess valid motor vehicle operator licenses are automatically fully appreciative of the need for observing continuing safety precautions in their driving and equipment operation. An intensive, continuing driving safety program should be maintained or at least supported by the utility manager, and each driving or equipment operating accident, regardless of how minor, should be subject to thorough investigation to seek out and correct bad driving and equipment operation habits. If employees are observed speeding, failing to stop at stop signs, or driving on the wrong side of roads on treatment plant sites, they should be subjected to disciplinary action. When these off-the-road violations are ignored, the violator will gradually tend to extend those unsafe driving practices to the public street and highway system, with eventual serious or even tragic accident results.

Within the treatment facilities of water and wastewater utilities, as in industries which use stationary process equipment, employees are continually exposed to numerous and severe injury hazards. Many accidents occur as the direct result of carelessness on the part of employees. Such carelessness can result in clothing or hair being caught in moving equipment, in the inadvertent leaning or walking into moving equipment, or the inadvertent placing of hands or feet in close proximity to moving belts, wheels, motors, and other equipment parts. These careless accidents obviously can cause injuries ranging from relatively minor bruises or cuts to severe fractures, loss of limbs, or even death. An intensive accident prevention program must be maintained among the employees of the water or wastewater utility, with emphasis placed on the development of employee awareness of potential hazards. It is also imperative that the utility manager be continually alert to the need for installing and maintaining fences, guards, railings, and other means to physically keep employees and other persons away from operating equipment and to alert them to dangerous conditions around the equipment. An additional safety precaution which must be observed around equipment is the positive locking out of controls which must remain in the "off" position and which must not be actuated under any circumstances while workmen are inspecting, lubricating, repairing, or otherwise working around such equipment.

Lifting accidents and falls annually result in a large number of injuries to employees of water and wastewater utilities, some of them serious lost-time injuries. Many lifting accidents result from lack of knowledge or gross lack of concern about lifting properly, with use of the legs rather than the back. Back injuries have for many years constituted major accident and injury problems for utility managers, and extensive indoctrination of employees as well as the providing and requiring the use of proper lifting equipment can help reduce the incidence of this type of accident. Several types of equipment are available for lifting, turning, and moving heavy barrels; pulling posts and stakes; lifting heavy loads to high elevations; and moving various sizes of equipment vertically and horizontally. It is the responsibility of the utility manager to be aware of potential hazards and to insist that supervisory personnel make the appropriate lifting and moving equipment available to employees who need it. Falls of many kinds result in injuries to employees throughout treatment plants and pumping stations, in the vicinity of excavated trenches, and also on and about various types of equipment. Some falls can be eliminated with the use of safety railings, enclosed ladders, and protected stairways, but if injuries from falls are to be reduced to virtual elimination, a program of indoctrination of employees on being adequately alert to the hazards of falls must be provided by the utility manager.

Electrical safety hazards of many kinds can be found in water and wastewater treatment plants, pumping stations, and other similar facilities which are operated and maintained by utility personnel. Electrical hazards range from serious accidents involving high-voltage electrical equipment and the accompanying massive and complex electrical switching gear, which provide many safety hazards, to simple electrical shock accidents resulting from use of a frayed or bare extension cord. An effective safety program will prevent the operation of electrical equipment and switching gear when it is being maintained or repaired, but such prevention will require that rigidly enforced procedures be implemented. Such procedures will include the physical locking of the switch gear and the labeling of the locked switch gear with appropriate signs or labels as well. Standard electrical maintenance procedures must require that all electric extension cords, cables, and electrical equipment which may have bare wires or other short circuit possibilities must be replaced immediately after inspection (which should be conducted on a routine basis by supervisory personnel and the safety officer) or casual observation by any employee has revealed deficiencies. The development, implementation, and maintenance of an adequate safety program which will reduce electrical hazards to an acceptable minimum will obviously require extensive investigative, developmental, and managerial efforts. Special emphasis must be placed on the design and location of electrical equipment and electrical systems. Operational and maintenance procedures will necessarily include numerous references to prohibited activities as well as detailed methods of starting, operating, and stopping equipment, switching from one electrical system to another, and providing both preventive and corrective maintenance to electrical equipment and systems. In addition to using the services of safety personnel of the water or wastewater utility for developing safe procedures for electrical equipment and systems, the manager of the utility should ensure that adequate input is also provided by the utility operations, maintenance, and engineering personnel as well as by other experienced personnel of consulting engineers, industry equipment manufacturers and suppliers, and the federal or state occupational safety and health regulatory agencies.

Serious personnel injuries in water and wastewater treatment plants often occur as the result of accidents involving chemicals used in the treatment processes. Injuries from chemicals can range from mild skin burns due to lime, ferric chloride, and cleaning chemicals to major eye damage caused by lime or other chemicals and even serious lung damage or death, which can result from inhalation of chlorine gas. Chemical burns to the hands, face, arms, and legs can result from improper handling of dry chemicals during the transfer from transport vehicles or railway cars to storage, during the transfer from chemical storage facilities to chemical feeder equipment, or because of malfunctioning of feeder equipment or improper operation of the chemical feeder equipment. Burns also often result from the use of chemicals in areas which are not provided with adequate dust collection equipment or area ventilation or because of faulty operation of such equipment. Burns from the use of liquid chemicals such as ferric chloride can result from careless or improper handling in transferring the chemical from the transport vehicle or railway car, in batch preparation of the liquid chemical, or during the feeding of the liquid chemical into the treatment stream.

One of the most serious types of chemical accident is the escape of chlorine gas from the railway tank car or chlorine gas cylinder, from the piping or chlorine evaporator, or from a chlorinator which has suffered some type of damage. Chlorine is an extremely toxic gas, and serious injury can result to persons if they inhale appreciable amounts. Minor exposure to chlorine gas will result in severe irritation to the eyes, nose, or throat, and significant exposure will normally result in lung irritation or lung damage. Prolonged exposure can lead to death. Essential safety precautions for water and wastewater utilities that have personnel working with or around chlorine gas include intensive safety training of employees, the wide use of warning signs, the providing and mandated use of special breathing apparatus under prescribed conditions, and rigid requirements for adherence to all acceptable chlorine safety precautions. Effective chlorine safety training courses are available from suppliers of liquid chlorine and from other sources, and the use of these training courses in the safety training programs of a water or wastewater utility can significantly improve the effectiveness of the total safety program of the utility.

Throughout water and wastewater treatment plants and pumping stations serious safety hazards can result from a failure to properly clean, inspect, and maintain equipment to keep it safely operable. This is particularly true of blowers or similar ventilation equipment in work areas where fumes which may be hazardous to worker's health must be eliminated. Another important consideration is the reliability of pumps which are needed for the pumping of water or wastewater from basins, galleries, and conduits in which or below which workers or equipment may be located. Any equipment which is not maintained in a clean, operable condition can malfunction, with resulting electrical shock hazards or hazards from physical injury caused by broken equipment. It should be obvious that at each facility there will be a need for a continuing program not only to maintain equipment but also to improve the preventive maintenance and housekeeping programs of the respective utilities.

Supervisors should be alert at all times to their responsibility to ensure that workmen always have the proper tools at hand for their specific work, and that those required tools are actually being used. Any hand tools which are worn to the point where they can be dangerous should be discarded. Hammers with rounded head surfaces, wrenches and screw drivers worn to the point where they do not properly fit nuts, bolts, and screws, and other similarly worn tools are all safety hazards and must be replaced with suitable safe tools.

The safety precautions which should be observed to protect men working in trenches to maintain or repair water or sewer facilities are largely comparable to those which are required for workmen who are engaged in the construction of water or sewer facilities. Emergency repairs of water main breaks or sewer breaks usually must be accomplished in extremely wet and muddy trench conditions, and this type of condition increases the safety hazards to workmen, especially if the trenches are deep. Under these circumstances, it is imperative that the excavations be made large enough to ensure adequate working space in the excavation and to minimize the possibility of workmen being trapped by trench cave-ins. In all situations in which shoring of trench walls is needed for adequate worker safety, work should be prohibited in the trench until adequate shoring is properly in place. It is also essential that enough dewatering equipment of the appropriate type be used on the job to ensure that the water in the excavation is held to an absolute minimum. Under wet trench conditions, it should be obvious also that electric cables and electric equipment should not be in close proximity to the workmen, who may be working in water. Another important safety precaution to be observed around trenches and other excavations is the total prohibition of the storing or parking of materials, tools, or equipment on the surface of the ground near the excavation, because of the continuing danger of such loads, whether heavy or light, causing trench cave-ins.

One of the more serious safety hazards which can be found in many sanitary sewer manholes is caused by the presence of sewer gas, which is odorless and difficult to detect, can cause workmen to lose consciousness in a few minutes, and can cause death in a relatively short period of time thereafter. Because of the lethal nature of sewer gas it is imperative that personnel who perform work in sanitary sewer manholes do so in accordance with rigidly enforced safety rules. These rules should require the wearing of safety harness equipment by any workmen who enter manholes, should require the venting of the manhole before a workman enters it, and also should make it mandatory that two or more persons work together so that one or preferably two men are available to pull a workman from a manhole if sewer gas causes him to lose consciousness. Safety rules should also require workmen to use appropriate equipment to check the air quality in the manhole before anyone enters. Because of the potential danger resulting from flammable gas collecting in sanitary sewer system lift stations and manholes, there is almost always a danger of explosion in the event that a spark is caused by electrical equipment or by a metal tool striking a manhole cover or manhole step, or because of carelessness on the part of a workman who has a lighted match, cigarette, or other flame in the vicinity of the lift station or manhole. These concerns and safety precautions should be included in the published and enforced safety rules of the utility.

A particularly serious safety hazard which requires special consideration and attention of the wastewater utility manager and employees within wastewater treatment plants is the digester gas which is produced by the anaerobic digestion of sludge from the wastewater treatment process. Digester gas normally is about 60 percent methane and as such is highly flammable and can be explosive. Extreme safety precautions must be exercised in the vicinity of any anaerobic sludge digester, or any pipe, equipment, or waste gas flare which may be connected in any way with the anaerobic sludge digestion process. Smoking of tobacco in any form and open flames of all types must be prohibited in the vicinity of areas where digester gas may be present, and explosion-proof electrical fixtures must be installed for use in the vicinity of those facilities which may be used for the storage, transmission, or use of digester gas. Special training of personnel in the use and handling of digester gas and working around digester gas is imperative. The manager of a wastewater treatment facility must ensure that all personnel who will at any time be in the vicinity of digester gas understand its flammable and explosive nature and how they must work and conduct themselves in its vicinity. Highly visible warning signs which dramatically and emphatically announce the danger of digester gas must be permanently and prominently displayed wherever digester gas can be expected to be present. These signs also should be in sufficient numbers to ensure that everybody near the digester gas will know of the probable presence of the gas as well as the absolute prohibition of smoking and open flames in such areas.

Safety Program Responsibilities

The manager of a water or wastewater utility is responsible for ensuring to the maximum extent feasible the safety of persons who work for the utility or who visit any of the facilities of the utility. The utility manager is also responsible for the protection of the property of the utility and any other property which may be near the facilities or activities of the utility. To meet these responsibilities the manager cannot rely on any uncoordinated safety activities or safety attitudes of the employees of the utility, but rather must develop a dynamic formalized safety program. This formalized safety program must result in specific safety procedures and rules which will be rigidly enforced. It will include extensive safety training of personnel in all of the operational activities which present serious safety hazards, and will designate individual responsibilities for safety inspection, safety training, safety meetings, and all of the other necessary aspects of an effective safety program.

The responsibility for developing and implementing an effective safety program within a utility obviously lies with the manager of the utility. The most effective manner in which the utility manager can fulfill this safety program responsibility is to ensure that each individual employee of the utility is aware of the importance of safety, is aware of the safety hazards involved with his work, and thinks about safety on a continuing basis. No safety program can really be effective unless the individual employees are willing and anxious to be safe and are enthusiastic about and supportive of a total safety program.

The development of proper employee attitudes toward safety and the elimination of safety hazards normally can be achieved only through and as the result of an organized formal safety program as mentioned previously. Such a safety program must be the direct responsibility of and under the direct supervision of a key individual on the manager's staff. This individual should be charged with the mandate to develop and manage a safety program which will involve the full support of all the supervisors within the utility. The safety program also should develop within each supervisor an enthusiastic and supportive attitude toward all aspects of safety. Through the supervisors each employee working group can be organized into intensive safety-oriented crews who will not only develop good personal safety habits, but will also insist on safe habits and safe working activities by other employees and will also demand an effective safety program to be maintained by the utility. Routinely scheduled safety meetings of both small and large groups of employees can be used effectively to concentrate the attention of the employee group on specific safety hazards which appear within the organization or to identify and call to the attention of employees any potential safety hazards which can and should be avoided or eliminated.

The safety program should be assigned and entrusted to the direct supervision of a person who is officially designated as the utility safety officer and whose sole duties are related to safety, especially in large utilities. In small utilities which cannot afford a full-time safety officer, at least the primary duties of an appropriate person should be related to safety and the development and implementation of a safety program. Even though to some utility managers it may appear that the salary of a full-time safety officer is extravagant, it has been found that an effective safety officer, through his efforts to reduce the number and severity of accidents within the utility organization, will provide benefits far in excess of his salary and the other costs of an effective safety program. The safety officer should normally be responsible directly to the manager of the water or wastewater utility and should be delegated specific authority, such as the authority to order unsafe activities halted and then to prescribe safe procedures under which the activities may continue. Such action by the safety officer obviously must not interfere with proper operations or maintenance activities of the utility to the extent that emergency conditions are created or to the extent that laws are violated.

No safety program of a water or wastewater utility will be effective or even meaningful unless the manager of the utility has a firm commitment to safety and the established safety program. This firm commitment should be communicated to all utility personnel so that everyone in the organization fully understands that safety is a high-priority concern of the manager and the higher levels of management of the utility. Occasions will arise when the safety officer will require the support of the utility manager in placing safety and accident prevention ahead of short-range operational or maintenance success. At such times the success of the entire safety program will depend on the decisions and actions of the utility manager. Individual employees of the utility will tend to develop the same commitment to safety and the safety program of the water or wastewater utility that they observe in the utility manager. It is thus imperative that the utility manager openly exhibit his full and unequivocal support for the safety program of the utility and the activities and authority of the safety officer.

The appointment of a safety committee and the enthusiastic support of its activities by the utility manager should help to ensure continuing support of a utility's safety program by its employees. This will be particularly true if employees from most of the major utility activities or departments are appointed to the safety committee and such service on the safety committee is widely publicized throughout the utility organization. The activities of a safety committee vary greatly, but some of the more important ones are the thorough review of accident investigations conducted by the safety officer, the consideration of safety suggestions by employees, the development of improved safety training programs and safety publicity activities, and the formulation of the agendas for the safety meetings of employees.

Regularly scheduled safety meetings for various groups of water or wastewater utility employees will enable the safety committee, the safety officer, the utility manager, and employees of the utility to develop the type of safety program which can be truly effective in reducing the number and severity of accidents experienced by the utility. The personal participation of individual employees is of utmost importance in making the safety meetings effective, meaningful, and valuable. The assignment of individual employees to the preparation for and conduct of scheduled safety meetings will enable individual employees to become more actively and enthusiastically involved in safety and the safety program of the utility. The rotation of safety meeting leadership among the employees will also tend to avoid the problem of employees being faced by the same person repeatedly speaking about safety. This also will enable individual employees to present views as part of the group rather than always as group leader with prepared safety information. Much better interaction between employees can thus be achieved by rotation of safety meeting leadership than would otherwise be possible. The improved safety meeting interaction between employees likewise should help to develop safer employees.

Many techniques are available to the safety officer and safety committee for emphasizing the importance of safety to employees and for increasing the concern of employees about safety. The use of safety posters, safety slogans, and safety messages throughout the communications system of the water or wastewater utility can be an effective technique. The use of these and other communication methods must be well planned, however, so that individual employees will not be so overwhelmed by safety messages that they lose interest completely in the safety program of the utility.

Many industrial and utility organizations have found that the sponsoring of safety competitions between and within groups of employees can be effective in improving accident prevention and employee safety attitudes and records. By using the pride that employees normally feel for their worker group the utility manager can expect that peer pressure will be exerted by such employee groups against individual employees who may be guilty of unsafe working practices or a complete disregard of specific requirements in the safety program of the organization. The manager of a water or wastewater utility should give serious consideration to sponsoring of such safety competitions as well as to furnishing some type of reward or prize for the group of employees who may win a safety competition.

The successes enjoyed by the National Safety Council and other safety organizations in the prevention of accidents and the improvement of safety in organizations of almost every kind makes it desirable for the water or wastewater utility manager to ensure that his utility is affiliated with and able to benefit from the expertise of such safety organizations. These established organizations have many years of experience in the development of effective safety programs and the prevention or reduction of many types of accidents. The use of the expertise of the personnel of safety organizations is yet another way in which the utility manager can effectively expand the knowledge and expertise available to his organization.

The institution of a special safety training program within a water or wastewater utility can reap huge benefits in accident prevention. Written and verbal instructions to employees concerning required safety measures cannot be nearly as effective as actual on-the-job training in specific safety precautions. Some examples of the many useful types of safety training include the proper handling of liquid and gaseous chlorine; the handling and storage of various potentially dangerous chemicals; the proper operation of valves, electric motors, and other equipment; the safe methods of joining and splicing pipe, conduit, and electric cable; safe driving training; and the proper use of tools and equipment.

Employee training in first aid should be another of the safety program efforts of the utility manager and the utility safety officer. Many accidents will occur, and fast first aid action by employees can often lessen the severity of the resulting injuries.

Summary

Because of the high cost of accidents in terms of financial loss, lost personnel time, and personal suffering, it is imperative that the manager of a water or wastewater utility establish a high priority for safety within his organization and that he provide personal leadership in that safety effort in the form of an effective safety program which he vigorously supports.

The savings to be realized from an effective safety program can be expected to be much greater than the costs of such a program, and it is thus imperative that the utility manager involve every employee of the utility in the development and maintenance of a continuing effective safety program.

An important part of an effective safety program is the person in the position of safety officer for the utility. This individual must have the full support of the utility manager, and must have the delegated authority to force compliance with safety requirements by employees of the utility. The safety officer should be responsible for the conduct of accident investigations, the development of the safety program of the utility, safety committee activities, and safety training.

15

Conservation and Resource Recovery

Over a period of many years, especially during the 1970s, most citizens of the United States came to the realization that their nation faced serious shortages of many resources not only regionally, but on a nationwide scale, and that some of the shortages were even critical. Even though substitutes will be found or developed for many resources which will be in short supply, conservation and the development of new resources will require major consideration in the United States throughout the foreseeable future.

Of particular concern to the managers of water and wastewater utilities are the shortages of fuel and other forms of energy, chemicals, certain building materials, and water of suitable quality for potable use or in some cases even suitable for industrial process use.

Shortages of certain natural resources will result in a scarcity of some construction materials and also of some components of the pumps, motors, valves, and other equipment required for the operation of water and waste-water facilities. These shortages can delay construction, result in inferior construction, lead to shortages of operable equipment with a corresponding problem in adequate treatment, and cause many other related problems.

Shortages of any of a wide range of resources can be expected to cause serious difficulties in scheduling and performing adequate maintenance of the equipment and facilities which must remain operable if water and wastewater utilities are to provide satisfactory, uninterrupted service to their customers. Inadequate construction materials or substandard equipment components will cause early deterioration or failure of facilities or equipment, thus increasing the amount of maintenance and repair work required within the utility. In addition, shortages of certain materials and replacement parts can delay or even prevent the proper maintenance of facilities and equipment with the result that certain treatment or pumping facilities or equipment will not continue to operate as they should. Some of the obvious results of these problems are shortages of potable water, lack of adequate water pressure, inadequate treatment of wastewater, and pollution of streams and the environment.

Operational constraints which can result from shortages of chemicals, fuel, electric power, natural gas, or other essential resources should be obvious. The continuing operation of water and wastewater treatment facilities without interruption requires that personnel, equipment, chemicals, fuel, energy, and the other necessities of treatment systems and treatment processes all be available where and when needed on a continuing basis at all times. When shortages prevent the acquisition and storage of adequate quantities of fuel, repair parts, or chemicals, interruptions in treatment with serious consequences can be expected.

The need for continuing and effective protection of public health and the environment as well as compliance with the legal requirements of federal and state regulatory agencies make it mandatory that all of the operations and maintenance activities of water and wastewater utilities remain uninterrupted. This obviously requires that all of the resources required of a water or wastewater utility must be provided on a continuing, uninterrupted basis. This is also true of the many other activities which are essential to a society which has become dependent on a continuing supply of energy and the products of industry and agriculture. Toward this end, it is imperative that major conservation and resource recovery activities be developed, especially within water and wastewater utility organizations, to ensure that resources will be available when needed to these agencies and to the general public.

The Need for Conservation of Energy, Water, and Other Resources

At times of serious shortages of natural resources, especially energy resources, whether on a local level, nationwide, or worldwide, it is imperative that the essential use of energy and other resources continue without interruption. Water and wastewater utilities provide services which are essential to the general public and must be provided on an uninterrupted basis. Uninterrupted service must be given a high priority. It is thus necessary for utility managers to develop the capability of assuring reliable supplies of the necessary energy and other resources for the use of the utility regardless of the magnitude or type of supply problems. It is also necessary for the utility manager to find ways of reducing the quantities of resources which are required for the operations of the utility without interfering with the essential operations of the utility.

If resources shortages of any kind result in or tend to cause a reduction in water supply quantity, or if a lowering of the level of wastewater treatment is mandated by such shortages, it will be necessary for the lifestyle of the affected people to change and perhaps change drastically. Such changes in lifestyle are extremely difficult or even impossible for many people to accept unless they are faced with a crisis which they both recognize and understand.

Potential serious energy shortages have been the concern of many knowledgeable governmental and industrial officials for a considerable number of years in the recent past, and until adequate acceptable alternative sources of energy have been developed it will be necessary for major conservation programs and major individual conservation efforts to be adopted by the American people, by the leaders of the industrial and commercial communities, and certainly by the managers and governing bodies of water and wastewater utilities.

The energy needs for which future supplies probably will not be adequate include petroleum fuels, electric power, and natural gas. Petroleum fuels still remain the major form of fuel for powering motor vehicles and heavy equipment, and with many of the activities of water and wastewater utilities dependent on the use of such motor vehicles and heavy equipment, it is imperative that these fuels be conserved to the maximum extent possible. A significant movement toward the use of other forms of energy for motor vehicles and heavy equipment is also warranted.

Numerous occurrences of power shortages and "brown outs" have plagued various sections of the United States during recent years, and in addition to a need for the development of additional sources of electric power it is essential that major conservation efforts be exerted throughout the nation and in all segments of society to reduce the demand on electric power generating facilities. The natural gas shortage in parts of the United States has resulted in curtailment of natural gas use by industry and by water and wastewater utilities, to the extent that some activities which are dependent on natural gas have been severely limited or completely halted. The conservation of natural gas along with the conservation of the other forms of energy must be an important part of water and wastewater facilities design and construction as well as the operations and maintenance procedures which govern the performance of utility personnel.

In several areas of the United States severe water shortages have mandated the development of major water conservation programs as well as the rigid curtailment of water uses. In such areas, when water shortages either exist or are anticipated it is obviously necessary for water conservation programs to be implemented. If the water conservation program is to be effective it must be well publicized, and a concerted public information effort by the water and wastewater utility officials will be necessary to ensure the cooperation of the entire community toward water conservation. In those states where people may rarely or never experience water shortages, it also is important that water conservation programs be developed for the purpose of conserving not only water, but also energy. Large quantities of energy are required to supply, treat, and distribute potable water; to transport and treat the wastewater which results from the use of domestic water; to produce and transport the chemicals needed for water and wastewater treatment; to heat the water used in homes and by commercial and industrial establishments; and to produce and transport construction materials and equipment needed to build new or expanded water or wastewater treatment facilities.

An important part of an energy conservation program for a water or wastewater utility is the continuing investigation of possible ways to make operational changes which can reduce the use of chemicals in the treatment process. Such reductions in turn will result in the conservation of the energy required to procure the raw materials from which the commercial chemicals are produced, in a reduction of the large quantities of energy required to produce the chemicals, and in a reduction of the fuel and energy required to transport the chemicals to the point of use.

The widespread conservation of water on a continuing basis will result in the postponement of the dates by which water or wastewater facilities must be expanded and thus will delay the time by which certain scarce construction materials and the materials which are used in the production of treatment equipment will be needed. This indirect effect of a water conservation program together with the benefits of energy conservation and cost reduction should be explained to the consuming public as a part of the water conservation public information program.

Resource Conservation Opportunities

Effective and significant conservation of resources by personnel of water or wastewater utilities or the general public normally can result only from conservation programs which direct the primary conservation efforts at those areas in which it is known that the results of resource conservation will be substantial. To direct the major conservation efforts at activities which can yield only minor savings rather than to search out activities which can bring major conservation results will obviously not produce an effective conservation program. This is not to dispute or to discount the belief that every bit helps, but only to encourage primary emphasis on activities which have the potential of bringing major benefits.

Some of the better known conservation programs in recent years have been water conservation programs which were necessitated by critical public water supply shortages. Many water conservation programs have been unsuccessful because prevailing public attitudes have not been supportive of sacrificial or even stringent water conservation efforts. Mandatory water conservation programs which would include the curtailment of lawn irrigation and other outside uses of water can be effective, especially if the penalties for violations are severe, but voluntary water conservation programs which require a high degree of understanding, concern, and support by the general public have usually been failures. It is particularly important to the success of a water conservation program for an effective public education effort to reach each segment of the general public to help people throughout the community to develop the attitude that they not only are willing, but are actually anxious to conserve water wherever possible in their daily lives. They will thus change the manner in which they shave, shower, use toilet facilities, wash dishes and clothes, and otherwise use water. They also will search out leaky faucets, pipe fittings and appliances to reduce the waste of water to an absolute minimum. The education of the public must be successful in developing an awareness of the comprehensive nature of the water conservation program, especially an understanding of the full range of resources being conserved along with the water. The public should understand that water conservation reduces energy requirements for treatment and distribution of water, reduces the use of raw materials, reduces the energy needs for the production of chemicals used in the treatment process, reduces the amount of resources required for treating as wastewater the water which is discharged to sewers, reduces the energy required for heating water for use in the home, and has other desirable impacts on the resources whose use can and should be reduced.

The conservation of petroleum fuels has been particularly difficult in the United States because the American citizen has become dependent on the flexibility and independence to be found in driving a motor vehicle whenever and wherever he wishes to be transported. This dependence on individual driving both for personal and business purposes will probably require a considerable future change in the transportation attitudes in both uses. A concerted mental effort, including a spirit of self-sacrifice by the motoring public, will be needed if the required changes in citizen attitude are to be realized. These driving attitudes are present among employees of utilities, including management personnel, and special conservation efforts are required of utility managers.

One of the successful methods of conserving petroleum fuels available to water and wastewater utility managers is a vehicle purchasing procedure which requires the specifying, purchase, and use of vehicles which have low gasoline or diesel fuel consumption ratings and records. This obviously will require some sacrifice of convenience, comfort, or power, but many individuals, commercial firms, and utilities have committed themselves to this transformation to lower-fuel-consumption vehicles. The conscientious continuing efforts by utility managers and employees of the utility to have vehicle engines tuned for maximum fuel efficiency at all times also can be effective in conserving fuel. Other fuel conservation measures such as prohibiting the idling of engines when vehicles are parked and the encouragement of shortened warm-up periods for engines obviously should also be considered and established.

Of all the fuel conservation methods available to the public, the most effective one is probably the use of car pools or mass transit. It has been estimated that if as few as 25 percent of the American work force were to carpool or use mass transit facilities to and from work a minimum of thirty million gallons of gasoline could be saved each workday. This is a significant result for a conservation effort which in most cases would provide only minimum inconvenience to those who would use the mass transit or car pool program. The water or wastewater utility manager should develop and give strong support to a car pool program for his employees and also encourage the use of mass transit to the maximum extent possible by his employees.

The conservation possibilities in the use of electric power are many and range from major electric power reductions in treatment processes to minor reductions in electric power for illumination. Each type of indoor and outdoor lighting fixture should be investigated for replacement of standard lamps and luminaires with new electricity-saving lamps or luminaires which can result in electric power consumption reductions of as much as 30 percent or more. Each lighted area should be studied for the purpose of determining the minimum amount of illumination needed for that area for purposes of safety, eyesight protection, and security. Illumination which is found to be unnecessary should be eliminated. Additional electric power conservation can be achieved by the installation of special devices which will not only monitor electric power surges in electrical equipment, but will also provide for major reductions in electric power consumption during periods of relatively low service demands on the equipment.

Most of the major opportunities for conservation of electric power will be found in studying and modifying water or wastewater treatment processes or in adjusting flows and loadings in such a way as to utilize as much as possible those treatment processes which require lower power consumption per unit of treatment than others. An example would be maximizing the use of high-purity oxygen aeration units in the secondary treatment process of a wastewater treatment plant rather than diffused air aeration units, which require from 20 to 35 percent more electric power per ton of biochemical oxygen demand reduction than do high-purity oxygen units. Efforts to reduce the power demand factor for electric power so as to reduce the extremely high cost of the power demand factor will not only be economical for the water or wastewater utility, but can also result in a reduction in the actual generating demand on electric power generation facilities. This obviously will be accompanied by a resultant increase in electric power available during periods of peak demand for other important uses in the community.

The consumption of natural gas or fuel oil for heating, incineration, and similar purposes can be reduced considerably by installing improved insulation for walls, ceilings, and windows to prevent as much heat loss as possible; by ensuring that doors are kept closed as much as possible during both the heating and air conditioning seasons; by developing affirmative employee attitudes toward energy conservation; and by providing for a less fuel-consumptive method than incineration for disposing of sewage sludge, trash, or other waste materials. The use of digester gas from the anaerobic digestion of sludge in wastewater treatment plants will be covered to a considerable degree in the resource recovery programs section of this chapter, but should be mentioned here as a means of conserving natural gas or fuel oil for heating or incineration purposes by substituting digester gas for natural gas or fuel oil in these activities.

A useful and important part of an energy conservation program is an energy audit. This type of investigation searches out and identifies the many specific uses of energy of various types throughout a building or installation, and quantifies the energy use for each segment of each activity within the facility. The findings of an energy audit can then be used as the basis for investigating in detail the major uses of energy, to eliminate those energy-wasting activities which are not an important part of the water or wastewater utility mission, and to reduce energy use systematically without disrupting operations in those functions which are essential to the water or wastewater utility.

For an energy conservation program within a water or wastewater utility organization to be effective it is essential that wholehearted support be obtained from virtually every employee. This degree of employee support can be realized only if attitudinal changes are accomplished by means of such techniques as extensive indoctrination of employees concerning the seriousness of energy shortages and the importance of individual employee conservation in the total energy conservation efforts of the utility and of the nation. The type of energy conservation information to be provided to employees should include data concerning the international and national relationships between energy production and energy consumption; the impact on the national economy of the dependence of the United States on other countries for energy sources; and the importance of energy conservation in the comprehensive program of balancing national energy consumption with the present and anticipated production of energy within the nation. A more meaningful and more personal aspect of employee indoctrination for effective energy conservation would be the development within employees of an awareness of the importance of each individual citizen's or employee's action taken to either waste or conserve energy. Employees should be provided with specific directives which set forth required actions for energy conservation. Some of these required actions are turning off lights not in use, closing doors to save heat or reduce air cooling, warming up vehicles for short periods of time rather than for long periods of time, conforming to vehicular speed limits, and avoiding the waste of hot water. Employees should also be made to understand the total implications of wasteful habits such as wasting hot water; they should be made to understand that resources such as hot water include not only the energy required to produce the potable water, but also energy needs for pumping and heating the water. Many other examples of individual employee opportunities to conserve energy can be cited to help employees develop an energy conservation attitude. Some employees may respond positively only if the emphasis is placed on the economics of energy conservation and the fact that as more money is wasted on energy consumption less money will be available for employee wages and other benefits.

For an energy conservation program to be effective, it is important that every employee of the water or wastewater utility should fully understand the importance of each individual effort in energy conservation. The phrase "every little bit helps" should be impressed on employees so that each employee will be concerned about such things as temperatures in buildings, the elimination of the waste of electric power used for unnecessary lighting, and the waste of petroleum fuels in warming up vehicles or operating vehicles in an improper and energy-wasteful manner. An intensive employee indoctrination program can result in a significant improvement in employee attitudes toward the energy conservation efforts of the utility and in the degree of success of the energy conservation program itself.

Examples of Resource Conservation

Many examples can be cited of successes in reducing water consumption. A plastic insert in the ½-inch-diameter shower arm can reduce maximum flow from the shower by 3 to 6 gallons per minute. Special inserted equipment in tank toilets can save up to 4 gallons per flush. Improved water-saving toilets can save from 2 to 4 gallons per flush. Improved water-saving shower heads can save from 5 to 8 gallons per minute, and improved water-saving automatic clothes washers and dishwashers can save from 5 to 15 gallons per load. The conscientious effort to use clothes washers and dishwashers only with full loads rather than partial loads also will have significant results in water conservation. Efforts on the part of water and wastewater utility managers to gain public acceptance of these conservation activities will have obvious positive impacts on the utilities.

One major wastewater utility found that the electric power required for each ton of biochemical oxygen demand removal by computer-controlled high-purity oxygen activated sludge treatment was 31 percent less than the power required by conventional diffused-air activated sludge treatment in the same plant. In addition to the benefits gained in energy conservation the utility also gained considerable economic benefit in reducing their $150,000 per month electric power bill.

These are but a few of the many examples which can be found of effective, beneficial resource conservation efforts.

Resource Recovery Programs

In addition to the responsibility for conservation of energy and other resources, the managers of water and wastewater utilities also have the responsibility to develop resource recovery activities within their utilities. Such resource recovery should include not only major efforts such as utilization of the gas from anaerobic digestion of wastewater sludge, but also relatively minor efforts such as the salvage of scrap metal, batteries, and other waste products.

One of the most valuable and readily available resources to be found in a water or wastewater utility is the digester gas which is produced in relatively large quantities as one of the byproducts of the anaerobic digestion of wastewater treatment plant sludge. Digester gas can be used beneficially in many ways. It can be used to provide heating of the digester contents; for providing the gas mixing of the contents of the anaerobic digesters; for the production of electric power; as a fuel for the direct drive power for pumps, blowers, and other equipment; and also as a fuel for heating other boilers or to be sold after being cleaned and processed. The use of digester gas as a fuel for motor vehicles shows promise, as demonstrated by several communities. Inasmuch as digester gas can be produced in significant quantities and on a relatively reliable basis, a comprehensive study should be undertaken by managers of wastewater utilities to determine the ways in which the gas should be used to provide maximum use and to provide maximum financial benefit to the utility customers and the general public.

For many years in the United States and around the world the sludge from wastewater treatment plants has been recovered and used as a valuable resource in the form of a soil conditioner and fertilizer. The most familiar example is Milorganite, which has been produced in Milwaukee for many years. Federal restrictions on the use of sludge obviously could reduce or eliminate this type of beneficial use of a valuable resource. Sludge has also been used, although only in limited applications, as a fuel and as a cattle and poultry feed. The major use of wastewater treatment plant sludge has been by farmers, who have used the sludge as a soil conditioner and fertilizer for agricultural purposes. An example of crop yield increase is shown in Figure 15-1, where the production of hay was increased by the prudent, controlled use of wastewater treatment plant sludge. The nutrient value of wastewater sludge is not so important as the organic nature of the sludge, but the enhance-

Figure 15-1 / Crop yield increase from the judicious application of wastewater treatment plant sludge.

ment of growth of lawns, agricultural crops, trees, bushes, and gardens has been well known and documented for many years.

Possible public health hazards which would result from improper use of unstable wastewater sludge have been a major concern in recent years, but there appear to be no recorded cases of illness or health problems as the result of the use of well established wastewater sludge for normal agricultural or lawn soil conditioning and fertilizer purposes. It is obviously essential that when wastewater treatment plant sludge is used in any way which will place it in contact with people or animals or within the food chain for animals or humans, the sludge must be organically stable and free from hazardous levels of toxic materials or other pollutants which would be health hazards. Certain federal agencies and groups of environmental reactionaries have not only voiced concern, but have vigorously protested the use of wastewater treatment plant sludge as a soil conditioner and fertilizer. Even though history has not supported these concerns or protestations, it should be obvious that considerable research and demonstration work will be required to prove the value of wastewater treatment plant sludge for agricultural, lawn, landscaping, and other similar purposes, and to prove the safety of the use of this type of sludge for these and other uses under adequately controlled conditions. The use of wastewater treatment plant sludge as a fuel has not been widespread, but the heating value of the sludge and the need for alternative sources of fuel make this a possible fuel source worthy of consideration and investigative research. The use of wastewater treatment plant sludge as a feed supplement for cattle feed and poultry feed or as a source of vitamins for both animal and human use likewise deserves serious consideration and research effort.

The water treatment process does not have the resource recovery possibilities which are present in the wastewater treatment process, but the recovery of treatment chemicals and backwash water are important resource recovery activities. The recovery and reslaking of lime for reuse in the water treatment process has been a common resource recovery activity in some water utilities for many years. When a resource recovery procedure such as the recovery and reslaking of lime is considered for use in a water treatment plant, the decision to reslake and use the lime is based on an evaluation of the costs and financial benefits resulting from the recovery and reslaking procedure. With energy conservation and energy production now important parts of utility management, these evaluations should determine both the financial and energy costs and benefits, and then the decision should be based on a combined financial and energy cost–benefit analysis rather than a financial analysis alone. Consideration should also be given to other resource recovery possibilities, such as the use of backwash water residue as an aid to coagulation and sedimentation in the water treatment process.

Resource recovery programs of water and wastewater utilities, in addition to those involving the major activities relating to treatment processes, should include the collection and recovery of metals and other waste materials of various types including barrels, automobile and truck batteries, pipe fittings, and broken equipment components. The financial salvage value of these waste materials may be relatively small or insignificant, but the recovery of these and other similar resources for reuse is important in the total resource recovery needs of the nation.

Broken pavement and building materials such as brick, block, and stone should be stockpiled and made available for use, for example, as riprap along stream channels for protecting treatment plant areas or sewers and water mains which are located across or adjacent to streams. Other fill materials which can be used for the protection of these and similar facilities or which can be used for other purposes should also be transported to conveniently located storage areas for future use.

One additional water and wastewater utility activity in which resource conservation and recovery can be practiced successfully is in the replacement of vehicles and other equipment. An effective equipment management program will normally include a continuing evaluation of the total annual costs of operation, maintenance, and replacement of vehicles and other equipment from an economic standpoint. In view of the importance of resource conservation, it is a responsibility of managers of water and wastewater utilities to provide the same type of evaluation from a resource conservation standpoint. If the equipment maintenance personnel can be reasonably expected to keep a vehicle or piece of equipment in operation for an additional period of time with only slightly increased annual maintenance costs, consideration should be given to continuing the use of existing equipment and vehicles for the longer period of time rather than replacing the vehicles or equipment.

Summary

The troublesome shortages of energy and other resources which have already been experienced and the expected future critical shortages of many national natural resources make it essential that managers of water and wastewater utilities develop, implement, and forcefully maintain vigorous programs of resource conservation and recovery within their utility organizations.

It is essential that water and wastewater utility managers develop programs to reduce the use of energy, particularly petroleum fuels, natural gas, and electric power in any way they can, consistent with adequate operation of the utility. Water conservation programs, likewise, are important to all water and wastewater utility operations because of the large amounts of natural resources and energy required for the supply, transmission, treatment, and distribution of water and the collection, pumping, treatment, and sludge processing and disposal activities of wastewater utilities.

Many resource and energy conservation possibilities exist within water and wastewater utility facilities, and it is of extreme importance that all employees of water and wastewater utilities be personally involved in conservation programs to reduce the use of fossil fuels, electric power, natural gas, and other resources.

Resource recovery programs within water and wastewater utilities can include many methods of recovering resources which can be recycled. In addition to the many opportunities for minor resource recovery in wastewater utilities, major opportunities are available for resource recovery in the form of digester gas from the anaerobic digestion of wastewater sludge, and also in the form of sludge from wastewater treatment facilities, for use as soil conditioners and fertilizer, and for fuel purposes.

16

Research and Development

During the 1970s the United States Congress enacted far-reaching water and wastewater legislation which required expanded technology in treatment, monitoring and sampling, chemical and biological analyses, and facilities design. The basic treatment technology in both water treatment and wastewater treatment had been largely unchanged for decades, but the new federal legislation in these fields required significant changes in levels and types of treatment, some of which could not be satisfied with known technology.

Concern throughout the nation for certain allegedly harmful chemical compounds and toxic organic substances in public water supplies and in wastewater sludge brought about a need to detect, measure, and provide adequate removal of compounds which previously were unknown and in many cases not measureable in the low concentration prescribed by federal law and the resulting regulations. Much of this concern was related to heavy metals and toxic or pathogenic substances which may be found in the effluent from wastewater treatment plants. These substances could possibly provide health hazards in public water supplies or in the sludge which resulted from the treatment of wastewaters and which could ultimately cause problems with groundwater pollution or with plant or animal uptake into the human food chain.

Concern for the protection of public health, environmental factors, politics, economics, and energy considerations all have a significant effect on present and future treatment technology in the water and wastewater fields. The impact of any or all of these factors at any specific future time will require revisions of and advancements in water or wastewater treatment technology, and it is imperative that each new or revised technology be developed, demonstrated, and appropriately applied by water or wastewater utilities at an early date and with a minimum of delay.

The Need for Research

The public scare caused by the identification of possible cancer-causing compounds in drinking water was largely responsible for the enactment by Congress of the Safe Drinking Water Act. The implementation of the Safe Drinking Water Act imposed on water utility managers the requirement for the removal from public water supplies of additional contaminants which had not previously been detectable or in some cases even known to be of concern. The removal of these contaminants and the monitoring of the concentrations and removal of the contaminants obviously would require the development of techniques for detecting and identifying the contaminants, the development of appropriate treatment techniques, and the development of adequate techniques for measuring the concentrations of such contaminants in the range considered to be dangerous to human health.

All of the pollutant removal techniques which would be required to bring compliance with the provisions of the new Safe Drinking Water Act had to be accomplished with reasonable assurance of adequate and reliable pollutant removals and at reasonably low cost. Detection and monitoring techniques obviously would be required before it could be determined for and within any water treatment facility that the required pollutant removals actually were being provided by the recognized and utilized treatment technology. As amendments to the Safe Drinking Water Act or the implementing regulations are adopted it will be necessary for a continuing program of research to keep pace with the need for improved detection, monitoring, and removal of contaminants.

Long before the enactment of the Safe Drinking Water Act, and also since that act became law, the Federal Water Pollution Control Act has gone through a number of major changes. In addition, many state laws pertaining to protection of public health and the environment have been enacted and amended. Most of these statutory changes have resulted in more stringent requirements for limiting the concentrations of pollutants in wastewater treatment plant effluent, for providing advanced wastewater treatment for some facilities, and for giving special consideration to new methods for the processing and monitoring of sludges from wastewater treatment plants, especially for those treatment agencies in which the sludge is to be beneficially used in the production of food as opposed to being destroyed.

It should be apparent that federal and state legislative and regulatory actions have made it necessary that additional basic and applied research efforts be exerted in order to develop new and improved methods of detecting, identifying, measuring, removing, and monitoring those pollutants considered or actually found to be harmful to the public health or the environment. It is reasonable to assume that future legislative and regulatory action on both the federal and state levels will likewise result in the need for additional continuing research activities.

Research is urgently needed into new methods of conserving energy and particularly into treatment processes which will utilize less energy than the processes which have historically been used. The use of high-purity oxygen aeration systems in wastewater biological treatment facilities has been found to require less electric power per unit of treatment than the usual diffused air facilities. Likewise, certain aeration basin diffusers have been found to be more efficient than others in oxygen transfer. A continuing research effort in these operational areas should certainly be beneficial throughout the nation's wastewater utilities.

Considerable research also is needed for developing sludge processing, disposal, and recycling systems which require less energy than do currently accepted sludge processing, disposal, and recycling systems. The increased use of waste heat for maintaining sludge digestion temperatures and for heating treatment facilities, the substitution of air-drying facilities for facilities which employ fuel-burning for sludge drying, and the development of other sludge processing or transportation methods which consume less energy than present methods are all examples of the types of research required for wastewater sludge handling. There is also a need for the development of water and wastewater treatment processes which will not only require less electric power in the treatment processes, but will also require smaller quantities of chemicals, will require chemicals which need less energy in their production, or can in any other way reduce the amount of energy consumed by water and wastewater utilities in their treatment processes.

The desire or necessity to provide coordinated or joint water and wastewater treatment for the purpose of either water reuse or the successive use of water has opened a wide range of research possibilities and needs. This needed research will hopefully result in the development and acceptance by regulatory agencies of processes which will combine the treatment of wastewater to a quality level which is acceptable for discharge to a waterway with the treatment of wastewater to such a level that the effluent from a wastewater treatment plant may be further treated for use in industrial water supply systems, for use in irrigation systems or for ultimate use as a portion of the potable water in potable water systems. Major savings of money and energy should be realized from reducing either the level of treatment in the wastewater treatment plant because of subsequent treatment to be provided in a water treatment process or from combining certain wastewater and water treatment processes into a common treatment process. Research is needed for the development and demonstration of the feasibility of these systems. Major research also is needed for developing faster and more economical methods of monitoring the influent and effluent at such water reclamation facilities for the full range of contaminants which may be of concern at present and in future years.

The development of a new industrial product has usually been accompanied by the production of new liquid wastes containing contaminants which have to be removed from the liquid waste streams from the particular industries. The treatment processes which historically have been employed in water and wastewater treatment facilities and which have generally been found to be satisfactory for many decades are now often found to be incapable of removing certain of these new contaminants. A considerable amount of continuing research is needed for the development of methods of adequately identifying and monitoring these many new contaminants and also for the development and perfecting of processes which will adequately remove these contaminants from the industrial waste streams. The research required by water and wastewater utilities therefore must include continuing investigation of the specific waste streams from the nation's industries, must be based on a continuing surveillance of these industrial waste streams, and must result in the development of adequate methods of treating such waste streams.

These are only a few of the situations and actions which make research necessary. The water or wastewater utility manager must be aware of the many causes of research needs, and must be prepared to participate in the needed research or at least help to ensure that the research will be funded and will be accomplished in a timely manner.

Responsibilities for Research

The question of who has the primary responsibility or any responsibility for the funding and conducting of the research required by water and wastewater utilities so that they can provide satisfactory service to the public cannot easily be answered. The local operating agencies responsible for providing either water service or wastewater service to the public, namely, water and wastewater utilities, have perhaps the greatest interest in having the results of appropriate research available to them as soon as needed for use in their day-to-day operations. Without adequate research most operating agencies could easily find themselves violating requirements of the Clean Water Act or the Safe Drinking Water Act. One of the unfortunate aspects of local operating agencies is that it has historically been far more difficult for them to provide adequate funds to finance the required research than it has been for other levels of government and private business. It should be obvious that regardless of the source of research funding, the responsibility for the required research should largely rest with the water and wastewater utilities. The manager of a water or wastewater utility has an obligation to ensure that his agency accepts the responsibility for ensuring that the research needed by his utility is funded and conducted when needed.

Inasmuch as it was the United States Congress and the U.S. Environmental Protection Agency who placed stringent demands for monitoring and treatment on the managers and governing bodies of water and wastewater utilities, and because the research which is required if these demands are to be met is needed by most water and wastewater utilities throughout the nation, the Federal Government obviously should provide at least the major portion of the funds needed for conducting the required research for developing any needed new technology. The Federal Government should also assume the responsibility for whatever research can best be sponsored or conducted by its own agencies. Water and wastewater utility managers obviously must exert whatever effort is necessary to force the Federal Government to meet their responsibility for the funding and performance of appropriate research.

Basic research in water and wastewater treatment has been conducted by many universities for many years, and these universities and other similar educational institutions will continue to play a key role not only in the performance of scholarly research, but also in providing guidance in and valuable input to applied research and development activities for improved water and wastewater treatment technology. Many valuable research projects have been conducted under the guidance of a team composed of personnel from a university and personnel from a water or wastewater utility, with expertise, facilities and costs being shared according to the needs and resources of each. This type of cooperative research conducted jointly by a water or wastewater utility and a university, whether with or without financial resources of the Federal Government or private business, should be encouraged and is recommended whenever possible.

Manufacturers of equipment which is used in water and wastewater treatment plants and in other types of water and wastewater facilities routinely perform a considerable amount of research and development work on a continuing basis. A sizeable research program is vital to most industrial organizations in their efforts to sell their share of the equipment to water and wastewater utilities and to private industry which is involved in industrial waste treatment. Whenever new monitoring or treatment requirements are placed upon water or wastewater utilities the equipment manufacturers obviously should be involved in a major leadership capacity in the development of new treatment processes and the equipment to be used in the new treatment processes. The same type of responsibility for development of new and improved products rests with the producers and suppliers of chemicals and materials which are used in the operation and maintenance of existing or new facilities. It is essential that the manager of a water or wastewater utility maintain a close relationship with manufacturers to ensure that new processes, equipment, chemicals, and materials will be available to him when needed.

Joint funding for research by several water or wastewater utilities in a manner similar to the way in which the research sponsored by the American Public Works Association Research Foundation has been financed makes possible many types of research on a scale which probably could not otherwise be accomplished. This type of cooperative or joint activity is recommended and should be considered for most water and wastewater utilities, which by themselves could experience great difficulty in financing the entire cost of necessary research. The pooling of funds from many water or wastewater utilities on the basis of population served, annual budget, or almost any other criterion can provide significant sums of money, which may even be adequate for all the research and developmental work which is needed. Research projects which are jointly funded by several utilities are normally under the guidance of a steering committee composed of representatives of several or all of the contributing agencies. Upon completion of the research, the results are made available first to the participating agencies and then to other interested agencies.

Other technical associations such as the American Water Works Association and the Water Pollution Control Federation have research foundations or some other type of research activity within their organization, and these associations are in a position to provide the administration, expertise and funding for much of the research and development work required by water and wastewater utilities. The sponsoring of research by these types of technical association can provide a reasonable distribution of the financial burden for research work among water and wastewater utilities in a manner similar to that by which the American Public Works Association distributes the cost of various types of research among public works agencies through its Research Foundation.

Justification of Research Expenditures

The expenditure of local agency funds for research conducted by and within water and wastewater utility organizations must usually be justified by the utility manager on the grounds that the specific utility will receive benefits from the research which are at least equal to, and preferably greater than, the cost of the research work. The history of research expenditures, activities, and benefits within the industrial sector provides one of the best general types of justification of research. Industries almost since the beginning of the industrial age have found it necessary to expend considerable energy, expertise, time, and money toward the development of new products, new tools and equipment, improved materials, and improved methods of producing better and less costly products for the consuming public. When developing the justification for research expenditures the managers of water and wastewater utilities thus should refer to the history of particular industries in their specific as well as more general research efforts and successes.

The water or wastewater utility manager should anticipate the need to provide a substantial amount of information to justify each research and development project which he recommends to his governing body for approval and funding. The justification presentation must include a clear definition of the research end result needed, the reason or the need for the utility to accomplish this end result, a clear explanation of why existing technology or methods are not adequate to meet the desired end result, and a comparison of the costs of performing the research and development work with the monetary as well as other, intangible benefits to be derived from the research and development work.

In many instances the water or wastewater utility manager can refer directly to the legal requirements, in the form of laws and regulations, for the utility to provide a level of treatment to comply with water quality standards, drinking water standards, wastewater sludge contaminant concentration limitations, and other similar requirements. Even though it should be relatively easy for the governing body of a water or wastewater utility to understand the need to perform research and development work for the purpose of developing better or less expensive methods of complying with certain legal requirements, the utility manager should plan to spend whatever time is necessary to develop a clear and complete justification presentation for the recommended work.

In the preparation of factual information which is to be the basis of justification for research projects, the utility manager should prepare the documentation in such a way as to not only show the total costs and total benefits to the citizens served by the specific water or wastewater utility, but also the total costs and benefits to the general public within the community, in the region, and across the nation, as well as to the environment, so that a complete comparison of costs and savings or benefits can be understood by the governing body of the utility. Many individuals serving on water or wastewater utility governing bodies will feel that research work financed and performed by their utility should result in appropriate benefits which accrue directly to the utility and to the people served directly by the utility. It is important that the utility manager counter such attitudes with an explanation of how the benefits which are derived by other water and wastewater utilities and other agencies from research funded by those other utilities will actually result in many tangible and intangible benefits to the local utility as well. As research is performed throughout the water or wastewater field, most utilities will eventually make use of the new technology, processes, or equipment which result from the research.

Documentation of the justification of research should include not only cost savings, but also such benefits as improvement in performance and reliability, conservation of energy and other resources, improvement in management capability, and development of increased knowledge within the water or wastewater field.

It should be emphasized that each research project must be justified to the utility governing body and to any other agency or association which will be asked to help fund the research project. The comparison of estimated costs and benefits together with a clear documentation of the need for the research must be provided and presented in an understandable manner to any individuals or groups who will be asked to share the costs of a research project, so that the needed funding for such research and development projects will be forthcoming without unreasonable delay or conditions.

Financing of Research

As was discussed in the section on the responsibility for conducting research, the financing of research and development projects within the water and wastewater field should and will come from many different sources. The one source of research funding which has been difficult to develop over the years has been the local utility annual budget. As long as research and development work is required to enable local agencies to comply with changing federal and state regulations and standards, it is essential that the annual budgets of water and wastewater utilities include funds for financing research and development work. This research and development work can either be funded and performed completely by the specific water or wastewater utility, or it can be part of a larger research and development project in which the local water or wastewater utility is merely contributing funds and expertise as a part of a major regional or national research and development project.

Inasmuch as most of the research and development work which will be required of water and wastewater utilities is the result of congressional or federal regulatory requirements, and because such research and development work will provide benefits on a national basis, it is certainly appropriate for managers of water and wastewater utilities to demand that the major part of financing of research in the water and wastewater field should be provided by the Federal Government. This high level of federal funding of research which should have been provided has not been available as needed in the past, and it is improbable that an adequate level of federal support of suitable research will ever be provided except with unreasonable constraints, conditions, and limitations on the expenditure of funds. It is important, however, that managers of water and wastewater utilities be persistent and exert continuing efforts to bring about adequate federal funding of research and development work with a minimum of conditions and restrictions, so that the research required for providing satisfactory water and wastewater service to the American people can be realized and accomplished properly.

Financial support for research into the improvement of water and wastewater activities should be vigorously solicited by managers of water and wastewater utilities from the manufacturers of water and wastewater equipment. Most manufacturers finance and conduct research which is intended to develop equipment, chemicals, and materials which they can sell in the water and wastewater utility market, but funds from these manufacturers should also be solicited for the support of basic research and developmental work which may be required for the development of new treatment processes as well as new and improved methods of operating existing treatment facilities to provide levels of treatment which will meet future public health and environmental requirements.

Within the United States there are numerous foundations which, in addition to other programs, regularly provide financial support to research of many types. It is the responsibility of utility managers to solicit financial support from these foundations for the types of research and development work which are required by water and wastewater utilities. Philanthropic foundations in particular should be approached for funding of research which is intended to provide benefits to public health, the environment, and the lifestyle of the general public. Research grants from these foundations are particularly attractive because they usually do not include restrictive constraints and requirements such as those which normally accompany federal grants for research and which usually interfere with the conduct and reporting of the results of research projects.

The previously mentioned cooperative research program sponsored by the American Public Works Association through their Research Foundation is an excellent means whereby individual water and wastewater utility organizations can benefit from costly research without providing more than a small portion of the funding required to finance that research. Other professional associations have similar types of research organizations, and full use should be made of the research funding and performance capabilities of all these associations, especially those dealing specifically with water and wastewater activities.

For many years a large amount of research in water and wastewater treatment has been accomplished by or in cooperation with universities and other institutions of higher learning. Although most of this has been basic research, much of the research has resulted in an ultimate improvement in treatment process, equipment, or operational methods. Managers of water and wastewater utilities should therefore develop a continuing cooperative relationship with nearby university officials to ensure that the full research capabilities of the universities combined with the expertise, facilities, and funds of the utility are used for the research and development work needed in the water and wastewater field. It is particularly important to the water or wastewater manager to ensure that the funds provided to universities for research are used for projects resulting in improvements to operating agencies.

Summary

Present and future federal and state regulations which place increasingly rigid restrictions on water and wastewater utility managers in providing adequate facilities, treatment, and operation will require that research efforts be accelerated in the water and wastewater field.

Research is required for the development of better methods of identifying and quantifying pollutant concentrations and for the development of improved methods of providing the treatment levels and types necessary to meet the requirements of the Safe Drinking Water Act and the Clean Water Act, as well as additional legislative and regulatory requirements which will certainly continue to confront the managers of water and wastewater utilities.

The responsibility for financing and conducting research in the water and wastewater fields must be assumed by the operating agencies, the Federal Government, universities, manufacturers, and the associations of professional and technical water and wastewater personnel throughout the nation. The development of cooperative research activities and attitudes among these groups will provide improved research results at minimum cost, with minimum duplication of efforts and minimum delays in making the results of meaningful research available to those who need the new information and techniques which have been developed by the research.

Each individual research project should be clearly and completely justified to the governing bodies of operating agencies, universities, or associations as well as to appropriate federal agencies in such a way as to ensure that the decision-makers understand the need for the research, the total costs involved in the research work, and the presumed and probable benefits to be derived from that research. Inasmuch as a major portion of the water and wastewater utility research to be required in the foreseeable future will be the result of action by the Federal Government, the bulk of funding of needed research should be provided by the Federal Government but without constraints and limitations which will make the results of the research either meaningless or too late.

Index